CAMBRIDGE LIBRARY COLLECTION

Books of enduring scholarly value

Life Sciences

Until the nineteenth century, the various subjects now known as the life sciences were regarded either as arcane studies which had little impact on ordinary daily life, or as a genteel hobby for the leisured classes. The increasing academic rigour and systematisation brought to the study of botany, zoology and other disciplines, and their adoption in university curricula, are reflected in the books reissued in this series.

Herbals:
Their Origin and Evolution

Agnes Arber (1879–1960) was a prominent British botanist specialising in plant morphology and the history of botany. In 1946 she became the first female botanist to be elected a Fellow of the Royal Society. First published in 1912 and issued in an expanded second edition in 1938, this volume traces the history and development of printed herbals between 1470 and 1670. This two-hundred-year period was the most prolific for the publication of herbals, and significantly saw the emergence of botany as a scientific discipline within the study of natural history. Although Arber mentions the medical aspects of the herbal, her analysis remains focused on investigating herbals from a botanical view, with chapters devoted to the evolution of plant descriptions, classifications and illustrations. Her book remains the standard work on this subject. The text of this volume is taken from a 1953 reissue of the 1938 second edition.

Cambridge University Press has long been a pioneer in the reissuing of out-of-print titles from its own backlist, producing digital reprints of books that are still sought after by scholars and students but could not be reprinted economically using traditional technology. The Cambridge Library Collection extends this activity to a wider range of books which are still of importance to researchers and professionals, either for the source material they contain, or as landmarks in the history of their academic discipline.

Drawing from the world-renowned collections in the Cambridge University Library, and guided by the advice of experts in each subject area, Cambridge University Press is using state-of-the-art scanning machines in its own Printing House to capture the content of each book selected for inclusion. The files are processed to give a consistently clear, crisp image, and the books finished to the high quality standard for which the Press is recognised around the world. The latest print-on-demand technology ensures that the books will remain available indefinitely, and that orders for single or multiple copies can quickly be supplied.

The Cambridge Library Collection will bring back to life books of enduring scholarly value (including out-of-copyright works originally issued by other publishers) across a wide range of disciplines in the humanities and social sciences and in science and technology.

Herbals:
Their Origin and
Evolution

A Chapter in the History of Botany, 1470-1670

AGNES ARBER

CAMBRIDGE
UNIVERSITY PRESS

CAMBRIDGE UNIVERSITY PRESS

Cambridge, New York, Melbourne, Madrid, Cape Town, Singapore,
São Paolo, Delhi, Dubai, Tokyo, Mexico City

Published in the United States of America by Cambridge University Press, New York

www.cambridge.org
Information on this title: www.cambridge.org/9781108016711

© in this compilation Cambridge University Press 2010

This edition first published 1953
This digitally printed version 2010

ISBN 978-1-108-01671-1 Paperback

HERBALS

THEIR ORIGIN AND EVOLUTION

A CHAPTER IN THE HISTORY OF BOTANY
1470–1670

A physician using a herbal

[After a picture by Adrian van Ostade, 1665, *Der Arzt in seinem Studier-zimmer*, 855c, in the Kaiser-Friedrich-Museum, Berlin] *Reduced*

HERBALS

THEIR ORIGIN AND EVOLUTION

A CHAPTER IN THE HISTORY OF BOTANY

1470–1670

BY

AGNES ARBER

M.A., D.Sc., F.R.S., F.L.S.

SECOND EDITION

REWRITTEN AND ENLARGED

CAMBRIDGE

AT THE UNIVERSITY PRESS

1953

PUBLISHED BY
THE SYNDICS OF THE CAMBRIDGE UNIVERSITY PRESS
London Office: Bentley House, N.W.1
American Branch: New York
Agents for Canada, India, and Pakistan: Macmillan

First edition, 1912
Second edition, enlarged and re-set, 1938
Reprinted, 1953

First printed in Great Britain at the University Press, Cambridge
Reprinted by offset-litho by Bradford & Dickens

TO THE MEMORY OF

PREFACE TO THE SECOND EDITION

In recent years marked advances have been made in the study of the history of botany, and I am grateful to the Syndics of the Cambridge University Press for giving me the opportunity of revising this book in the light of newer knowledge. In the process of rewriting I have experienced, to the full, both how essential and how difficult it is to follow Flaubert's advice: "Quant aux corrections, avant d'en faire une seule, remédite l'ensemble." In the attempt to refashion the book into greater wholeness, I have been helped by my daughter Muriel's criticism.

I have kept the general plan of the volume unaltered, but I have repaired certain omissions in Chapter IV by adding sections dealing with botany in Spain and Portugal, and with the origin of herbaria. Now that the literature of the history of science has become a world in itself, I should have liked to give references, page by page, for all the facts or opinions which I owe to previous writers, but this would have involved increasing the size of the book unduly, and burdening the text with an apparatus of footnotes, which would have been oppressive to the general reader. I have, however, trebled the number of works mentioned in the list of sources (Appendix II), and I have indexed it under subjects (Appendix III); students will thus, I hope, be enabled to check and amplify my statements.

My study of herbals has now extended over so long a period, that to recall all those who have given me help would lengthen this introduction beyond measure; all that I can do is to supplement the acknowledgments in my former preface by alluding to the most outstanding of my present debts. In the first place, I must name Prof. Sir Albert C. Seward, F.R.S.; it was out of a suggestion of his that this book originally arose, and the second version owes no less to him. I should also like to express my gratitude to my friend, Gulielma Lister, for the stimulus of her encouragement and critical interest, without which this edition would not have come into being.

One cannot bear a subject in mind for a quarter of a century without discovering at the end of that time that some reorientation of one's outlook has taken place. In this connection I owe much to Charles and Dorothea Singer, since it is largely through countless discussions with them, and with the biologists who foregather in their hospitable library, that I have come to see the history of the herbal in juster perspective. Amongst others who have given me invaluable help, and saved me from many errors, I must mention Dr T. A. Sprague, Deputy Keeper of the Herbarium and Library, Royal Botanic Gardens, Kew, who has allowed me to draw repeatedly upon his expert knowledge both of modern botany, and of botany in the sixteenth century; and Dr A. C. Klebs, of Nyon, to whom I have been able to appeal on points connected with the incunabula. I wish also to express my gratitude to Mr A. F. Scholfield, Cambridge University Librarian, for giving me every assistance in my work; and to Bodley's Librarian, Oxford, as well as to the authorities of the Preussische Staatsbibliothek, Berlin, and of the Bibliothèque Nationale, Paris, for the courtesy with which they have answered my enquiries. It is a pleasure, moreover, to thank Prof. F. T. Brooks, F.R.S., who has permitted me to use the herbals in the Botany School, Cambridge; Prof. W. Rytz, of Bern, who has helped me by the communication of his discoveries concerning Weiditz; Dr B. Milt, of Zurich, who has been so kind as to give me unpublished information about Gesner; and Mr W. T. Stearn, of the Lindley Library, Royal Horticultural Society, who has allowed me to take advantage of his special studies in the history of systematics. For aid in bibliographical matters, as well as facilities for examining books under their care, I am indebted to Mr H. M. Adams, Librarian of Trinity College, Cambridge; Mr J. Ardagh, Librarian of the Department of Botany, British Museum (Nat. Hist.); Mr H. R. Creswick, of the University Library, Cambridge; Mr G. R. Driver, Librarian of Magdalen College, Oxford; Mr H. Guppy, John Rylands Library; Drs M. Sabbe and Bouchery of Antwerp; Mr Spencer Savage, of the Linnean Society; Miss E. M. Smelt, Librarian of the Pharmaceutical Society; Dr H. Bramley Taylor, the Worshipful Society of Apothecaries; Dr B. E. J. Timmer, the Universiteits-Bibliotheek, Amsterdam; and Mr Aubrey F. G. Bell.

I am obliged to the Clarendon Press, Oxford, for permission to use a quotation from *The Turkish Letters of Ogier Ghiselin de Busbecq,* translated by Mr E. S. Forster; and to the Editors of the Loeb Classical Library, who have allowed me to quote from Sir Arthur Hort's translation of the *Enquiry into Plants* of Theophrastus.

Part of the work on the history of botany incorporated in the present edition has been carried out during the tenure of a Leverhulme Research Fellowship; I wish to acknowledge my indebtedness to the Trustees for the opportunities thus afforded me. I should also like to thank those reviewers and correspondents from whom, at the time of the first publication of this book, I received corrections and constructive suggestions, of which, after this long interval of time, I have been able to make use.

In connection with the new figures added to the present edition, I must express my gratitude to Dr F. W. T. Hunger, for allowing me to reproduce pictures from his facsimile version of *The Herbal of Pseudo-Apuleius*; to Prof. Gola, for giving me information about the portrait of Prospero Alpino at Padua, and for having it photographed for reproduction here; to Prof. R. E. Fries, for lending me an engraved portrait of Brunfels from the collection in the Hortus Bergianus at Stockholm; to Dr Wegener, for giving me a photograph of

Preface to the Second Edition

the painting of the dragon-tree in the Clusius Collection in the Preussische Staatsbibliothek at Berlin; to the Director of the Kaiser-Friedrich-Museum, Berlin, who has allowed me to reproduce Adrian van Ostade's *Physician in his study*, as a frontispiece to the present book; and to Les Archives Photographiques d'Art et d'Histoire, Paris, for sanctioning the use of François Clouet's portrait of Pierre Quthe, from the Musée du Louvre. For permission to copy illustrations from books in collections under their charge, I am indebted to the Cambridge University Librarian; to the Keeper of the Department of Printed Books, British Museum; and to the Director of the British Museum (Nat. Hist.).

I must, moreover, express my gratitude to Sir Henry Maxwell-Lyte, K.C.B., for his kindness in giving me information about his ancestor, Henry Lyte, and showing me family records relating to him.

I wish, also, to thank the members of the staff of the Cambridge University Press for the trouble which they have taken over the production of this edition; authors who benefit by their critical judgment and expert skill, owe more to them than it is easy to express.

Finally I should like to say that I am acutely conscious that the treatment of many topics in this book is of the slightest and sketchiest character. The attempt to deal with so large a subject as the botanical history of two centuries, within so small a compass as that of the present volume, resolves itself into an essay in the art of rejection. I can but echo the words with which William Turner sent forth his herbal, nearly four hundred years ago: "To them that compleyne of the shortnes of the boke, I answer, if they be...learned men, let them write longer bokes and amend my shortnes with their long and great bookes."

AGNES ARBER

CAMBRIDGE
January, 1938.

FROM THE PREFACE TO THE
FIRST EDITION, 1912

The main object of the present book is to trace in outline the
evolution of the *printed herbal* in Europe between the years
1470 and 1670, primarily from a botanical, and secondarily
from an artistic standpoint. The medical aspect, which could
only be dealt with satisfactorily by a specialist in that science,
I have practically left untouched, as also the gardening
literature of the period. Bibliographical information is not
given in detail, except in so far as it subserves the main
objects of the book. The titles of the principal botanical works,
which were published between 1470 and 1670, will be found
in Appendix I.

The book is founded mainly upon a study of the herbals
themselves. My attention was first directed to these works
by reading a copy of Lyte's translation of Dodoens' herbal,
which happened to come into my hands in 1894, and at once
aroused my interest in the subject. I have also drawn freely
upon the historical and critical literature dealing with the
period under consideration, to which full references will be
found in Appendix II. The materials for this work have
chiefly been obtained in the Printed Books Department of the
British Museum, but I have also made use of a number of
other libraries. I owe many thanks to Prof. Seward, F.R.S.,
who suggested that I should undertake this book, and gave
me special facilities for the study of the fine collection of
old botanical works in the Botany School, Cambridge. In
addition I must record my gratitude to the University
Librarian, Mr F. J. H. Jenkinson, M.A., and Mr C. E. Sayle,
M.A., of the Cambridge University Library, and also to
Dr Stapf, Keeper of the Kew Herbarium and Library. By the

kindness of Dr Norman Moore, Harveian Librarian to the Royal College of Physicians, I have had access to that splendid library, and my best thanks are due to him, and to the Assistant-Librarian, Mr Barlow. To the latter I am especially indebted for information on bibliographical points. I have also to thank Mr Knapman of the Pharmaceutical Society, Dr Molhuizen, Keeper of the Manuscripts, University Library, Leyden, and the Librarian of the Teyler Institute, Haarlem, for giving me opportunities for examining the books under their charge.

The great majority of the illustrations are reproduced from photographs taken directly from the originals by Mr W. Tams of Cambridge, to whom I am greatly indebted for the skill and care with which he has overcome the difficulties incidental to photographing from old books, the pages of which are so often wrinkled, discoloured or worm-eaten. For the use of pl. xx[1], which appeared in *Leonardo da Vinci's Note-Books*, I am under obligations to the author, Mr Edward McCurdy, M.A., and to Messrs Duckworth & Co. Text-figs. 7, 18, 89, 90 and 130 are reproduced by the courtesy of the Council of the Bibliographical Society, from papers by the late Dr Payne, to which the references will be found in Appendix II, while, for the use of text-fig. 126, I am indebted to the Royal Numismatic Society. For permission to utilise the modern facsimile of the famous Dioscorides manuscript of Anicia Juliana, from which pls. i, ii, xviii, and xxiii are derived, I have to thank Prof. Dr Josef Ritter von Karabacek, of the k. k. Hofbibliothek at Vienna. In connection with the portraits of herbalists here reproduced, I wish to acknowledge the generous assistance which I have received from Sir Sidney Colvin, formerly Keeper of Prints and Drawings, British Museum.

I would also record my thanks to Mr A. W. Pollard,

---

[1] The numbering of the plates and text-figures mentioned in this preface has been altered to that in the present edition.

From the Preface to the First Edition

Secretary of the Bibliographical Society, Prof. Killermann of Regensburg, Signorina Adelaide Marchi of Florence, Mr C. D. Sherborn of the British Museum (Natural History) and Dr B. Daydon Jackson, General Secretary of the Linnean Society, all of whom have kindly given me information of great value. For help in the translation of certain German and Latin texts, I am indebted to Mr E. G. Tucker, B.A., Mr A. F. Scholfield, M.A., and to my brother, Mr D. S. Robertson, M.A., Fellow of Trinity College, Cambridge.

I wish, further, to express my gratitude to my father for advice and suggestions. Without his help, I should scarcely have felt myself competent to discuss the subject from the artistic standpoint. To my husband, also, I owe many thanks for assistance in various directions, more particularly in criticising the manuscript, and in seeing the volume through the press.

AGNES ARBER

Balfour Laboratory, Cambridge,
26 July 1912

CONTENTS

Contents

xvi

LIST OF ILLUSTRATIONS

xvii

Plates

List of Illustrations

FIGURES IN THE TEXT[1]

[The initial letters, which will be found at the beginnings of the chapters, are taken from Pierre Belon's *Les Observations de plusieurs singularitez et choses memorables, trouvées en Grece, Asie, Judée, Egypte, Arabie, et autres pays estranges, ...Imprimé à Paris par Benoist Prevost. 1553.*]

[1] The dates refer, in each case, to the particular edition from which the figure has been copied, which is not always the first. For fuller titles and dates of first editions, see Appendix I.

Figures in the Text

XX

Figures in the Text

xxi

Figures in the Text

xxii

Figures in the Text

Figures in the Text

Chapter I

THE EARLY HISTORY OF BOTANY

1. INTRODUCTORY

In the present book, the special subject treated is the evolution of the *printed herbal*, between the years 1470 and 1670, but it is impossible to arrive at clear ideas on this subject without some knowledge of the earlier stages in the development of botany. The first chapter will therefore be devoted to the briefest possible sketch of the progress of botany before the invention of printing, in order that the position occupied by the herbal in the history of the science may be realised in its true relations.

From the beginning, the study of plants has been approached from two widely separated standpoints—the philosophical and the utilitarian. Regarded from the first point of view, botany stands upon its own merits as an integral branch of natural philosophy, whereas, from the second, it is merely a by-product of medicine or agriculture. At different periods in the evolution of the science, one or other aspect has predominated, but from classical times onwards, it is possible to trace the development of these two distinct lines of enquiry, which have, at happy moments, converged, though they have more often, to their detriment, followed unconnected routes.

In the western world, botany, as a branch of natural philosophy, may be said to have owed its inception to the

I. The Early History of Botany

unparalleled mental activity of the finest period of Greek culture. From this time onwards the nature and life of plants were brought within the scope of research and speculation, the results of which have been handed down to us.

2. ARISTOTELIAN BOTANY

Aristotle, Plato's pupil, who lived from 384 to 322 B.C., concerned himself with science in the broadest sense, and his influence in this field, especially during the middle ages, dominated European thought. The greater part of his botanical work is unfortunately lost to us, but his writings on other subjects include many references to plants, from which we can gather some of his general ideas. He held that each member of the world of living things had its "psyche"; for this expression, which cannot be Englished rightly by any single word or phrase, we may use "soul", or "vital principle", as an approximate translation. On Aristotle's view, the soul of the plant was "nutritive" only, and thus on a lower plane than the soul of movement and feeling in animals, and the reasoning soul in man. The long survival of these ideas may be witnessed by a quotation from Trevisa's version of the encyclopaedia of Bartholomaeus Anglicus, which was printed in 1498: "For trees meve [move] not wylfully fro place to place as beestes doo: nother chaunge appetitte and lykynge, nother felyth sorowe....In tres is soule of lyfe...but therin is no soule of felynge."

Aristotle left his library to his pupil Theophrastus (b. 370 B.C.), naming him as his successor. Theophrastus was well fitted to carry on the great traditions of the school, since he had, in earlier years, studied under Plato himself. We happen to know far more about the botany of Theophrastus than about that of Aristotle, since a work has come down to us, called the *Enquiry into Plants*,[1] which may perhaps have been compiled from the notes made by scholars who attended the lectures of

[1] The quotations from this book are taken from Sir Arthur Hort's translation; see Appendix II.

2

Theophrastus. The *Enquiry* opens with a discussion of the parts of plants, which the author tries to interpret by analogy with the organs of animals, though he points out that the correspondence is markedly imperfect. He realises the difficulty of fitting the vegetable world into any hard-and-fast scheme, and he concludes, rather wistfully: "In fact your plant is a thing various and manifold, and so it is difficult to describe in general terms." He proceeds to a classification of plants, which we shall consider in a later chapter (p. 163); but he is careful to point out that his proposed divisions are somewhat arbitrary, and that plants may pass from one class to another. The most striking quality of the writings of Theophrastus is, indeed, the way in which they combine delicate discrimination with freedom from dogmatism. In one passage, for instance, we find him at war with over-precise definitions, while, in another, he urges his readers not to neglect distinctions, even if they are admittedly not absolute.

The *Enquiry* is chiefly concerned with the plants of the Mediterranean region round about Greece, but it also shows some knowledge of the botany of other lands. It is believed that Theophrastus owed part of this knowledge of foreign plants to Alexander the Great, who also had been a pupil of Aristotle, and who was so much alive to the value of science that he took trained observers with him to the far east, to bring back reports on what they had seen. The parts of the *Enquiry* dealing with such subjects as "The plants of rivers, marshes, and lakes, especially in Egypt", or "The plants special to northern regions", are the earliest studies to shadow forth the ecological standpoint, which, in modern botany, has become of special importance.

Throughout the dark ages, the botany of the Aristotelian school was little known in western Europe, but in the thirteenth century it was revived through a surprisingly indirect channel. Soon after the time of Alexander, the foundation of Greek schools began in Syria. From these centres

3   1-2

the teachings of Aristotle were handed on into Persia, Arabia, and other countries. The Arabs translated the Syriac versions of Greek writers into their own language, and their physicians and philosophers kept alive the knowledge of science during the early mediaeval period, in which Greece and Rome had ceased to be the homes of learning, and while culture was still in its infancy in Germany, France, and England. The Arabic translations of classical writings were eventually rendered into Latin, and then, sometimes, even into Greek again, and in these guises found their way to western Europe.

Amongst other works which suffered these successive metamorphoses, was a pseudo-Aristotelian treatise, *De plantis*, which is now attributed to a certain Nicolaus Damascenus. We do not know when he was born, but his historical setting is indicated by the fact that Herod the Great sent him on a mission to Rome a few years before the birth of Christ. His book about plants is a compilation based primarily upon Aristotle and Theophrastus; an English version of it has recently appeared. It is of importance in the annals of western science, because it formed the starting-point for the botanical work of Albertus Magnus.

Albert of Bollstädt (d. 1280), Bishop of Ratisbon, was a famous scholastic philosopher. He was esteemed one of the most learned men of his age, and was called Albertus Magnus during his lifetime, the title being conferred on him by the unanimous consent of the schools. The Angelic Doctor, St Thomas Aquinas, became one of his pupils. His botanical work forms only a small fraction of his writings, but it is with that section alone that we are here concerned. It is embodied in a treatise, *De vegetabilibus*, dating from before A.D. 1256. Although Albertus undoubtedly found the framework for his botanical ideas in *De plantis*, which he revered as Aristotle's own words, he had too strong a mind to follow any authority slavishly, and no little part of what he wrote was original. He

Plate i

ϹΟΓΧΟϹ ΤΡΑΧΥϹ, *Sonchus* sp. [Dioscorides, *Codex Aniciae Julianae* (Vind. Med. Gr. I), circa A.D. 512, facsimile, 315 recto] *Reduced*

was in many ways in advance of his time, especially in the suggestions which he offers as to the classification of plants, and in his observations on detailed structure in certain flowers. We shall return to his writings in future chapters dealing with these subjects. It will suffice now to mention his remarkable instinct for morphology, in which he was probably unsurpassed during the next four hundred years. He points out, for instance, that, in the vine, a tendril sometimes occurs in place of a bunch of grapes, and from this he concludes that the tendril is to be interpreted as a bunch of grapes incompletely developed. He distinguishes also between thorns and prickles, and realises that the former are of the nature of stems, while the latter are merely a surface development.

Despite his insight into structure, Albertus had an unfortunate contempt for that branch of the science now known as systematic botany. He considered that to catalogue all the existing species was too vast and detailed a task, and one altogether unsuited to the philosopher. However, in his Sixth Book he so far belied his principles as to give descriptions of a number of plants.

Albertus was troubled with many subtle problems connected with the vegetable "psyche"; he questions, for instance, whether, in the material union of two individuals, such as the ivy and its supporting tree, their souls are also united. Like Theophrastus, and other early writers, Albertus held the theory that species were mutable, and illustrated this view by pointing out that cultivated plants might run wild and become degenerate, while wild plants might be domesticated. Some of his ideas, however, on the possibility of changes from one species to another, were quite baseless. He stated, for instance, that, if a wood of oak or beech were razed to the ground, an actual transformation took place, aspens and poplars springing up in place of the previously existing trees.

On the subject of the medicinal virtues of plants, the statements of Albertus, in their temperate tone, contrast favourably with the puerilities of many later writers. Much of the criticism from which he has suffered at various times has been, in reality, directed against a book called *Liber aggregationis*, or *De virtutibus herbarum*, of which he was supposed, erroneously, to be the author. We shall refer to this work again in Chapter VIII.

After the days of Albertus, no great development occurred in Aristotelian botany until the time of Andrea Cesalpino, whose writings, which belong to the end of the sixteenth century, will be considered in later chapters.

3. MEDICINAL BOTANY

Throughout its course, Aristotelian botany suffered from one serious handicap—an inadequate basis of actual fact. It came into existence at the time when Greek philosophy was at its height, and it owed its development to men who were completely at home with general ideas, but who were unaware that, before a theoretical treatment of the vegetable world was possible, it was necessary to know in detail what plants were really like, and how they lived. Such knowledge could not be deduced merely from general principles constructed by the human mind; it needed also minute and prolonged observation. No doubt the Aristotelian botanists would have been capable of such observation, but they were not alive to the necessity for it; it was left for workers in the, apparently, less promising field of medicine to lay the foundations of the copious and exact knowledge of plants which we possess to-day. From very early times a variety of herbs had been used as healing agents, and it had been necessary to study them in special detail, in order to discriminate the kinds employed for different purposes. It was from this purely utilitarian beginning that systematic botany for the most part originated. As we shall show in later chapters, very many of the herbalists whose work we have to discuss were medical men. Moreover, it is not taxonomy alone that we owe in the first instance to medicine. Nehemiah Grew (1641–1712), one of the founders of the science of plant anatomy, was led to embark upon this subject because his anatomical studies as a physician suggested to him that plants, like animals, probably possessed an internal structure worthy of investigation, since they were the work of the same Creator.

In all parts of the world systems of folk medicine have been developed, but we are here concerned with Greece alone, since it was from that region that herbal knowledge made its way into western Europe. In ancient Greece there was considerable traffic in medicinal plants. The herbalists[1] and druggists[2] who made a regular business of collecting, preparing, and selling them, do not appear, however, to have been held in good repute; Lucian makes Hercules address Aesculapius as "a root digger and a wandering quack". The herb gatherers evidently aimed at creating a monopoly by fencing their craft about with all manner of superstitions handed down by word of mouth, most of which had for their moral that herb collecting was too complicated and dangerous a pursuit for the uninitiated. With the *Enquiry into Plants* of Theophrastus, a Ninth Book is included, which is probably a compilation brought together at some date after the death of the reputed author. In this treatise certain of the herb gatherers' directions for collecting medicinal plants are quoted, though with ridicule. We learn that he who would obtain peony root was advised to dig it up at night, because, if he did the deed in the day-time, and was observed by a woodpecker, he risked the loss of his eyesight. The superstitions connected with procuring mandrake and black-hellebore are also cited with contempt. It seems that the herb collectors declared that "one should draw three

---

[1] ῥιζοτόμοι = root diggers, or herb gatherers, for ῥίζα signifies a medicinal plant in general, as well as a root.

[2] φαρμακοπῶλαι = drug sellers.

circles round mandrake with a sword, and cut it with one's face towards the west; and at the cutting of the second piece one should dance round the plant.... One should also, it is said, draw a circle round the black-hellebore ... and one should look out for an eagle both on the right and on the left; for that there is danger to those that cut, if your eagle should come near, that they may die within the year."

Though the Ninth Book of the *Enquiry* contains a good deal about medicinal herbs and their uses, it is of less importance in the history of botany than the work of a later Greek, Krateuas (Cratevas), physician to Mithridates, who began to reign in 120 B.C. The writings of Krateuas are no longer extant; we possess only fragments, embedded in the books of other writers. He produced, as we know from Pliny, a herbal containing coloured pictures of plants. The chief part of our knowledge of his work has reached us through one of his successors, Pedanios Dioskurides, who was born in Asia Minor, and whose life was passed probably in the first century of the Christian Era, in the time of Nero and Vespasian. He was a medical man, and he speaks of having seen many lands in his military travels, so it seems not unlikely that he was an army doctor.

Dioscorides—to give him the name by which he has generally been known in this country—compiled a work which is usually cited under its Latin title, *De materia medica libri quinque*; in this treatise he included about five hundred plants. No contemporary version has survived; the only manuscript which we shall consider here is Byzantine, and dates from about A.D. 512. It was made for Anicia Juliana, a noble lady whose father, Flavius Anicius Olybrius, had once been, for a brief space, Emperor of the West. Juliana, who lived into the age of Justinian, was renowned for her ardent Christian faith, and for the churches which she built. It is probable that the manuscript associated with her name remained in Constantinople during the first millenium of its

8

history. In 1562 we hear of it in a letter[1] written by the diplomatist Ogier Ghiselin de Busbecq, who had just returned from Turkey. "One treasure", he says, "I left behind in Constantinople, a manuscript of Dioscurides, extremely ancient and written in majuscules, with drawings of the plants and containing also, if I am not mistaken, some fragments of Cratevas.... It belongs to a Jew, the son of Hamon, who, while he was still alive, was physician to Soleiman. I should like to have bought it, but the price frightened me; for a hundred ducats was named, a sum which would suit the Emperor's purse better than mine. I shall not cease to urge the Emperor to ransom so noble an author..... The manuscript, owing to its age, is in a bad state, being externally so worm-eaten that scarcely any one, if he saw it lying in the road, would bother to pick it up."

About seven years later, the great Codex was conveyed to the Imperial Library in Vienna, having been purchased, either by the Emperor, or, more probably, by Busbecq himself. It is still to be seen in Vienna to-day,[2] while a facsimile reproduction has made it accessible to students in other countries. Examples of the figures which it contains are shown on a reduced scale, and without colour, in pls. i, f.p. 4; ii, f.p. 10; xviii, f.p. 186; xxiii, f.p. 240. We shall return to these pictures in Chapter VII.

The earliest versions of Dioscorides appear to have been unillustrated, and there is reason to believe that some, if not all, of the pictures in the Vienna Codex were ultimately derived from Krateuas. In the part of this manuscript in which the text is specifically attributed to him, nine kinds of plant are named. It is a striking sign of the continuity of botany that seven of these nine names should have survived into the nineteenth century, or later, as generic terms.

[1] The original letter is in Latin; it is quoted here from the translation in Forster, E. S. (1927); see Appendix II.

[2] This manuscript is described technically as *Codex Vindobonensis Med. Gr.* 1; among Dioscorides manuscripts it is known as *Constantinopolitanus*.

These ancient names include *Aristolochia, Anemone,* and *Anagallis.*

The text of *De materia medica* consists, in the main, of an account of the names and healing virtues of the herbs enumerated. The actual descriptions are very slight, and it is only those plants with particularly salient characters which can be recognised with any certainty. Nevertheless, the importance of the part played by this treatise can scarcely be overestimated. Up to the height of the renaissance period, and later, *De materia medica* was accepted as the almost infallible authority. As picturing the general attitude, we may quote from a contemporary Elizabethan translation of a book by a Spanish physician, Nicolas Monardes. He tells us that Dioscorides, whithersoever he went, "did seeke these Herbes, Trees, Plantes, Beastes and Mineralles, and many other thinges, of the whiche he made those sixe bookes, whiche are so celebrated in all the worlde, wherby he gate the glory and fame, whiche we see he hath, and there hath remained more fame of hym, by writing them, then although he had gotten many Cities with his warlike actes". Another indication of the way in which *De materia medica* was regarded in the sixteenth century, is the fact that, in 1568, William Turner referred to Luca Ghini—who was the first occupant of the chair of botany founded at Bologna in 1534—as "reder of Dioscorides in Bonony".

In the seventeenth century, the fame of Dioscorides was unabated. In 1633 Thomas Johnson wrote that *De materia medica* "is as it were the foundation and grounde-worke of all that hath been since delivered in this nature", and, at as late a date as 1652–5, John Goodyer thought it worth while to make an interlinear translation of the whole work, simply as an assistance in his own studies. Moreover, even to-day, botanists remain in touch with Dioscorides, for in Sibthorp's great *Flora Graeca,* which is still in steady use as a work of reference, the names given in *De materia medica* are often

Plate ii

ΣΤΡΑΤΙΩΤΗΣ Ο ΧΙΛΙΟΦΥΛΛΟΣ, *Ferula* sp. [Dioscorides, *Codex Aniciae Julianae* (Vind. Med. Gr. I), circa A.D. 512, facsimile, 322 verso] *Reduced*

cited side by side with those bestowed by Linnaeus and later authorities.

In the period which is our primary concern in the present book, the elucidation of Dioscorides was one of the chief preoccupations of many of the herbalists. Pierandrea Mattioli is the most renowned of these commentators; his work on the subject ran through more than fifty editions in various languages. Ruellius, Amatus Lusitanus, and many others, were also active in the study, while Luigi Anguillara deserves to be held in high honour as a commentator of unusual insight. Discussion of *De materia medica* was not indeed confined to the specialised writings of these scholars, but formed an integral part of most of the sixteenth-century herbals. The botanist of to-day may find it difficult to avoid a feeling of impatience at the amount of time and energy devoted, in earlier times, to the elucidation of Dioscorides; but this impatience is less than just. It is a truism that slavish dependence upon authority hampers the progress of science; but in the earlier history of botany this dependence had its value, since it led to the preservation of the text of Dioscorides through the dark and middle ages; and when the great awakening of the renaissance came, the struggle to understand his text led gradually to a knowledge of floras which hardly could have been obtained without the particular motive and clue thus supplied. It must be admitted that some of the feebler herbalists were completely paralysed by their veneration for Dioscorides, but those of original mind, who faced his work critically, and realised its limitations, were able to use it without allowing it to shackle them; Antonio Brasavola (1500–55), for instance, made it clear that, in his view, the herbs described by Dioscorides did not include a hundredth part of those that grew upon the earth.

Within its own geographical range, *De materia medica* has not fallen wholly into disuse, even in the twentieth century. In 1934 the Director of Kew, when visiting the Athos

11

peninsula, met an Official Botanist Monk. On his excursions in search of "simples", this functionary carried with him, in a bulky black bag, four volumes in manuscript, described as having been copied from Dioscorides. With the aid of these folios, he satisfied himself as to the names of his plants.

One of the contemporaries of Dioscorides, Gaius Plinius Secundus, commonly called the Elder Pliny, should perhaps be mentioned at this point, although he was not a physician, nor does he deserve the name of a philosopher. In the course of his *Natural History*, which is an encyclopaedic account of the knowledge of his time, he treats of the vegetable world. He refers to a far larger number of plants than Dioscorides, probably because the latter confined himself to those which were of importance from a medicinal point of view, whereas Pliny mentioned indiscriminately any plant to which he found a reference in any previous book. Pliny's work was chiefly of the nature of a compilation, and indeed it would scarcely be reasonable to expect much original observation of nature from a man who was so devoted to books, that it was recorded of him that he considered even a walk to be a waste of time; but he deserves his little niche in the history of botanical terminology, since, in describing the lily, he used the word "stamen" for the first time in the modern sense.

Many manuscript herbals were written in western Europe in the time between the classical period and the end of the fifteenth century. These works showed little originality; their authors were content to base them upon Greek and Latin writings and Arabic[1] commentaries. In this brief sketch of the history of botany before printing, we can scarcely do more than mention the existence of such manuscripts; in the two succeeding chapters something must be said, however, about those which stood in the closest relation to the first of the printed herbals.

---

[1] The botany of the Arabs deserves a detailed treatment, which cannot be given in the present book.

Chapter II

THE EARLIEST PRINTED HERBALS
(Fifteenth Century)

1. The Encyclopaedia of Bartholomaeus Anglicus and *The Book of Nature*

A fter the invention of printing with movable type in the middle years of the fifteenth century, a period of active book production followed, during which many works which previously had passed a lengthy existence in manuscript, were put into circulation in print, side by side with others which were then new. The result is that a number of the "incunabula"—the technical name for books printed before the year 1501, at the time when the craft was in its "swaddling clothes"—are in reality far more ancient than their dates of publication would suggest. This characteristic is illustrated in two of the earliest printed books containing strictly botanical information. One of these is the *Liber de proprietatibus rerum* of Bartholomaeus Anglicus, a monk, who is sometimes mistakenly identified with Bartholomew de Glanville (d. 1360). This work was first printed about 1470, and no less than twenty-five editions of it appeared before the end of the fifteenth century; the author was not, however, of that period, but was a contemporary of Albertus Magnus. One of the sections of his encyclopaedia consists of an account of a large number of trees and herbs, arranged in alphabetical order, and is chiefly occupied with their

13

medicinal properties. It also contains some theoretical considerations about plants on Aristotelian lines.

The other early printed book, which, though not a herbal, included botanical lore, was *Das pûch der natur* of Konrad, or Cûnrat, who is sometimes called von Megenberg. It came from the press of Hanns Bämler of Augsburg in 1475. Long before it was multiplied by printing, *The Book of Nature* had a wide circulation. A large number of manuscripts still exist, as many as eighteen being preserved in the Vienna Library and seventeen at Munich. The basis of the printed book was a German translation from a Latin original, which had been compiled in the thirteenth century by a pupil of Albertus. The portion dealing with plants consists of an account of the virtues of a number of trees, herbs, and vegetable products, with their Latin and German names. The chief interest of the work, from our present point of view, is that it contains the earliest botanical woodcuts known to exist, one of which is reproduced in pl. iii. We shall return to this subject in Chapter vii.

2. The *Herbarium* of Apuleius Platonicus

A herbal has been defined as a book containing the names and descriptions of herbs, or of plants in general, with their properties and virtues. The word is believed to have been derived from a mediaeval Latin adjective, "herbalis", the substantive "liber" being understood. It thus simply means "herb book". Among the first of the printed works to which the term "herbal" is commonly applied, is a little Latin treatise, the *Herbarium* of Apuleius Platonicus. The identity of the original compiler of this work is unknown, and, in order to distinguish him from the Apuleius of *The Golden Ass*, he is sometimes called Apuleius Barbarus, or Pseudo-Apuleius. His herbal is an illustrated medical recipe book, of small importance in comparison with the Dioscoridean *De materia medica*, from which, and from Pliny, it is in the main derived.

Plate iii

Woodcut of Plants [Konrad von Megenberg, *Das půch der natur*, 1475]
Reduced

The Herbarium *of Apuleius Platonicus*

Thomas Johnson (the seventeenth-century editor of Gerard's *Herball*) considered that it was probably written, in the first place, in Greek, and this opinion is still widely held. The treatise may date from the fifth century; if this is so, it must

Fig. 1. "Plantago", Plantain [*Herbarium Apuleii Platonici*, ?1481]

have had a career of a thousand years in manuscript before it first appeared in print at Rome.

It is conjectured that Johannes Philippus de Lignamine, who was responsible for the earliest printed edition, was a Sicilian; he is known to have been a member of the entourage of Pope Sixtus IV. He explains in his book that the manuscript on which he based his version came from Monte

15

II. Fifteenth-century Herbals

Cassino. Until a few years ago, it was assumed that this manuscript was no longer extant; but recently it has been suggested that a codex of the ninth century, still preserved at Monte Cassino, may possibly be the source of the printed *Herbarium*. This manuscript and the printed book have now been reproduced in facsimile, so arranged that they can be compared page by page. The resemblance of the illustrations is often striking, but there are also considerable differences. The two sets of pictures undoubtedly belong to the same family, but whether those in the printed book were derived directly from those in this particular codex remains, at present, an open question. The figures of the water-lily and saxifrage, as they appear in the manuscript and the printed book, are reproduced here (figs. 72 and 73, p. 167; 83 and 84, p. 187). Other illustrations from the printed version of the *Herbarium* are shown in pls. iv, f.p. 16; v, f.p. 18; xix, f.p. 188; and figs. 1, p. 15; 2, p. 17. In Chapter vii we shall have something more to say about these pictures, which must have been copied from manuscript to manuscript for hundreds of years before they made their début in print. Only one of their peculiarities need be mentioned here; it is that, if a herb is credited with the power of healing the bite or sting of an animal, that animal is shown with the plant in the same cut; for example, the plantain is accompanied by a serpent and a scorpion (fig. 1, p. 15). This device must have been for the information of those who could not read; it becomes less frequent in later herbals.

Soon after the appearance in Italy of the first printed edition of the *Herbarium* of Apuleius Platonicus, three works of great importance were published at Mainz in Germany These were the *Latin Herbarius* (1484), the *German Herbarius* (1485), and, derived from the latter, the *Hortus* or *Ortus sanitatis* (1491). The *Latin* and the *German Herbarius*, together with the *Herbarium* of Apuleius, may be regarded as the doyens among printed herbals. It is not improbable that

16

Plate iv

"Orbicularis" [*Herbarium Apuleii Platonici, ?1481*]
(*The tint represents contemporary colouring*)

all three were largely based upon pre-existing manuscripts representing a tradition of great antiquity, but it is only for the Apuleius that we have direct evidence of this.

HERBA ARTEMISIA LFPTA.
FILOS.I.MATRICALE.

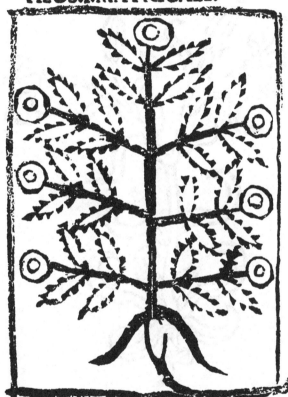

Fig. 2. "Artemisia", Wormwood [*Herbarium Apuleii Platonici,* ?1481]

The various forms of the *Latin* and the *German Herbarius*, and of the *Ortus sanitatis*, are described under many names, since there is usually no title-page, in the modern sense, to settle the matter. Indeed, if a herbalist of that date had wanted a specific title for his book, it would have been by no means easy for him to find an appropriate term, since botanical language was in its infancy. In mediaeval thought, analogy played so conspicuous a part, that it is not surprising to find treatises about plants called by such names as *Hortus* or *Gart*, in which use is made of the analogy of a garden. Even at as late a date as 1616, Olorinus, in his book about *Wonder*

Fig. 3. "Lilium" [*Latin Herbarius*, 1484]

Trees, speaks of them as transported "from the great world garden" ("Auss dem grossen Weltgarten") into "this little paper gardenlet" ("diss kleine Papieren Gårtlein"). The vagueness of such descriptions as *Hortus* and *Herbarius* makes the editions of the incunabula herbals difficult to unravel. Moreover, in those early days of the press, before the necessity for copyright had been realised, as soon as a popular

Plate v

"Mandragora", *M. officinalis* Mill., Mandrake [*Herbarium Apuleii Platonici*, ?1481] (*The tint represents contemporary colouring*)

work was published, pirated editions and translations sprang
into existence. In the case of the *German Herbarius*, a new
edition was printed at Augsburg only a few months after the
appearance of the original at Mainz. Such editions were often
undated, and the sources from which they were derived were
seldom acknowledged.

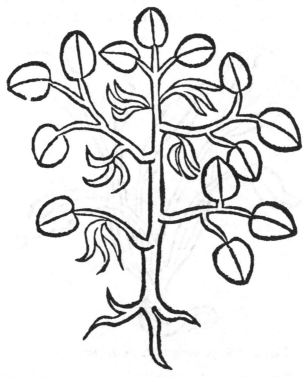

Fig. 4. "Aristologia longa" [*Latin Herbarius*, 1484]

The passage of the earliest printed books through the press
was inevitably a slow process, as compared with the rapid
production of the present day. The result was that the printer
had leisure to make occasional alterations, so that different
copies belonging actually to the same edition sometimes show
slight variations. The bibliographer has thus to deal with yet
another element of confusion.

2-2

II. Fifteenth-century Herbals

3. The *Latin Herbarius*

The work to which we may refer for convenience as the *Latin Herbarius* is also known under many other titles—*Herbarius in Latino*, *Aggregator de simplicibus*, *Herbarius Moguntinus*, *Herbarius Patavinus*, etc. In the form of a small quarto, it was first issued by Peter Schöffer in 1484 at Mainz. This town was one of the earliest homes of typography; it is

Fig. 5. "Serpentaria" [*Latin Herbarius*, 1484]

widely believed that printing with movable type originated there in the decade 1440–50.

Other early editions and translations of the *Herbarius* appeared in Bavaria, the Low Countries, Italy, and probably also in France. The work, like most of the early herbals, was anonymous, and was a compilation from mediaeval writers, and from certain classical and Arabic authors. It seems to have no connection with the *Herbarium* of Apuleius, which is nowhere cited. The majority of the authorities quoted wrote

20

before A.D. 1300, and no author is mentioned who might not have been known to a student in about the middle of the fourteenth century, that is to say, at least a hundred years before the *Herbarius* was published. This suggests that the work may have existed previously in manuscript form.

BRIONIA

Fig. 6. "Brionia" [*Latin Herbarius* (Arnaldus de Villa Nova, *Tractatus de virtutibus herbarum*), 1499]

The wood-blocks of the first edition printed in Germany are bold and decorative, but show little attempt at realism (figs. 3, p. 18; 4, p. 19; 5, p. 20; 85, p. 190). A different and more attractive set of figures was used in Italy to illustrate the text (figs. 6; 63, p. 150; 74, p. 169; 86, p. 191; 87, p. 192; 88, p. 194). The authorship of the version of the *Herbarius* from which these pictures are taken, has been attributed erroneously to Arnaldus de Villa Nova, a physician of the thirteenth century; this mistake arose because, on the first page of an earlier edition, there was a woodcut of Avicenna and Arnaldus de Villa Nova, whose names are quoted in the preface.

In the *Latin Herbarius*, the descriptions and figures of the herbs are arranged alphabetically. All the plants discussed were natives of Germany or in cultivation there, and the object of the work seems to have been to help the reader to the use of cheap and easily obtained remedies, in cases of illness or accident.

4. The *German Herbarius* and related works

Of even greater importance than the *Latin Herbarius*, is the *German Herbarius*, also called the *Herbarius zu Teutsch*, *Gart der Gesundheit*, the *German Ortus sanitatis*, the *Smaller Ortus*, or *Cube's Herbal*. This folio, which was the foundation of the later works named *Hortus (Ortus) sanitatis*, appeared at Mainz, also from the printing press of Peter Schöffer, in 1485, the year following the publication of the *Latin Herbarius*. It has been mistakenly regarded by some authors as a mere translation of that book. The *German Herbarius* seems, however, to be to a great extent an independent work. If the statements in the preface can be trusted, the originator of the treatise was a rich man, who had travelled in the east, and the medical portion was compiled under his direction by a physician. The latter was probably Dr Johann von Cube, who was town physician of Frankfort at the end of the fifteenth century.

The preface to the *Herbarius zu Teutsch* begins with the words, "Offt und vil habe ich by mir selbst betracht die wundersam werck des schepfers der natuer." Similar words are found in all the different German editions, and in the later *Hortus sanitatis* they are translated into Latin.

The medical ideas which this introduction presents are

based upon the theory of the *four elements*, and the *four principles* or *natures*. The theory had been maintained, but not originated, by Aristotle, and it held its own for a period of two thousand years. As an instance showing its prevalence in Shakespeare's day, we may recall Sir Toby's rhetorical question—"Do not our lives consist of the four elements?" Since the preface to the *Herbarius zu Teutsch* lucidly expounds the ideas connected with the *elements* and the *natures*, and also reveals, clearly and delightfully, the spirit in which the work was undertaken, it is translated[1] below, almost *in extenso*:

"Many a time and oft have I contemplated inwardly the wondrous works of the Creator of the universe: how in the beginning He formed the heavens and adorned them with goodly, shining stars, to which He gave power and might to influence everything under heaven. Also how He afterwards formed the four elements: fire, hot and dry—air, hot and moist—water, cold and moist—earth, dry and cold—and gave to each a nature of its own; and how after this the same Great Master of Nature made and formed herbs of many sorts and animals of all kinds, and last of all Man, the noblest of all created things. Thereupon I thought on the wondrous order which the Creator gave these same creatures of His, so that everything which has its being under heaven receives it from the stars, and keeps it by their help. I considered further how that in everything which arises, grows, lives or soars in the four elements named, be it metal, stone, herb or animal, the four natures of the elements—heat, cold, moistness and dryness—are mingled. It is also to be noted that the four natures in question are also mixed and blended in the human body in a measure and temperament suitable to the life and nature of man. While man keeps within this measure, proportion or temperament, he is strong and healthy, but as soon as he steps or falls beyond the temperament or measure of the four natures, which happens when heat takes the upper hand

[1] Translated from the second (Augsburg) edition of 1485 by E. G. Tucker.

and strives to stifle cold, or, on the contrary, when cold begins to suppress heat, or man becomes full of cold moisture, or again is deprived of the due measure of moisture, he falls of necessity into sickness, and draws nigh unto death. There are many causes of disturbances, such as I have mentioned, in the measure of the four elements which is essential to man's health and life. In some cases it is the poisonous and hidden influence of the heavens acting against man's nature, for from this arise impurity and poisoning of the air; in other cases the food and drink are unsuitable, or suitable but not taken in the right quantities, or at the right time. Of a truth I would as soon count thee the leaves on the trees, or the grains of sand in the sea, as the things which are the causes of a relapse from the temperament of the four natures, and a beginning of man's sickness. It is for this reason that so many thousands and thousands of perils and dangers beset man. He is not fully sure of his health or his life for one moment. While considering these matters, I also remembered how the Creator of Nature, Who has placed us amid such dangers, has mercifully provided us with a remedy, that is with all kinds of herbs, animals and other created things to which He has given power and might to restore, produce, give and temper the four natures mentioned above. One herb is heating, another is cooling, each after the degree of its nature and complexion. In the same manner many other created things on the earth and in the water preserve man's life, through the Creator of Nature. By virtue of these herbs and created things the sick man may recover the temperament of the four elements and the health of his body. Since, then, man can have no greater nor nobler treasure on earth than bodily health, I came to the conclusion that I could not perform any more honourable, useful or holy work or labour than to compile a book in which should be contained the virtue and nature of many herbs and other created things, together with their true colours and form, for the help of all the world and the common good. Thereupon I caused this praiseworthy work to be begun by a Master learned in physic, who, at my request, gathered into a book the virtue and nature of many herbs out of the acknowledged masters of physic, Galen, Avicenna, Serapio, Dioscorides, Pandectarius, Platearius and others. But when, in the process of the work, I turned to the drawing and depicting of the herbs, I marked that there are many precious herbs which do not grow here in these German lands, so that I could not draw them with their true colours and form, except from hearsay. Therefore I left unfinished the work which I had begun, and laid aside my pen, until such time as I had received grace and dispensation to visit the Holy Sepulchre, and also Mount Sinai, where the body of the Blessed Virgin, Saint Catherine, rests in peace. Then, in order that the noble work I had begun and left incomplete should not come to nought, and also that my journey should benefit not my soul alone, but the whole world, I took with me a painter ready of wit, and cunning and subtle of hand. And so we journeyed from Germany through Italy, Istria, and then by way of Slavonia or the Windisch land, Croatia, Albania, Dalmatia, Greece, Corfu, Morea, Candia, Rhodes and Cyprus to the Promised Land and the Holy City, Jerusalem, and thence through Arabia Minor to Mount Sinai, from Mount Sinai towards the Red Sea in the direction of Cairo, Babylonia, and also Alexandria in Egypt, whence I returned to Candia. In wandering through these kingdoms and lands, I diligently sought after the herbs there, and had them depicted and drawn, with their true colour and form. And after I had, by God's grace, returned to Germany and home, the great love which I bore this work impelled me to finish it, and now, with the help of God, it is accomplished. And this book is called in Latin, *Ortus Sanitatis*, and in German, *gart d'gesuntheyt* [garden of health]. In this garden are to be found the power and virtues of 435 plants and other created things, which serve for the health of man, and are commonly used in apothecaries' shops for medicine. Of these, about 350 appear here as they are, with their true colours and form. And, so that it might be useful to all the world, learned and unlearned, I had it compiled in the German tongue. * * * * * *

Now fare forth into all lands, thou noble and beautiful Garden, thou delight of the healthy, thou comfort and life of the sick. There is no man living who can fully declare thy use and thy fruit. I thank Thee, O Creator of heaven and earth, Who hast given power to the plants, and other created things contained in this book, that Thou hast granted me the grace to reveal this treasure, which until now has lain buried and hid from the sight of common men. To Thee be glory and honour, now and for ever. Amen."

The pictures in the *Herbarius zu Teutsch* are, on the whole, drawn with greater freedom and realism than those of the *Latin Herbarius*; these woodcuts—of which examples are shown in figs. 7, p. 27; 89, p. 195; and 90, p. 196—form the basis of nearly all the plant figures for the next half century, being copied and recopied from book to book. No work which excelled or even equalled them was produced, until a new period of botanical illustration began with the herbal of Brunfels, published in 1530.

Not only were the figures in the *German Herbarius* extensively used in later books; the text, also, was much copied, and translated into other languages. The rare early French herbal called *Arbolayre* was formerly described as one of these translations, but a more recent suggestion is that it is a faulty reproduction of some manuscript based upon a work by Platearius called *Circa instans*. Platearius' account of herbs held an important place in the centuries before the invention of printing, and a version of it is believed also to have formed the basis of another French herbal, *Le grand Herbier*. The latter work is of special interest to British botan-

Fig. 7. "Acorus", *Iris Pseudacorus* L., Yellow-flag [*Herbarius zu Teutsch*, Mainz, 1485]

ists, since it was translated into English and published in 1526 as *The grete herball*; this book, which also owes something to the *Ortus sanitatis*, will be considered at length in the next chapter.

5. THE *HORTUS* (*ORTUS*) *SANITATIS*

The third of the fundamental botanical works produced at Mainz towards the close of the fifteenth century, was the *Hortus*, or, as it is more commonly called, *Ortus sanitatis*,

Fig. 8. "Leopardus" [*Ortus sanitatis*, 1491]

printed by Jacob Meydenbach in 1491. It is in part a modified Latin translation of the *German Herbarius*, but it is not merely this, for the virtues of the herbs are dealt with at greater length, and it also contains treatises on animals, birds, fishes, and stones, which are almost unrepresented in the *Herbarius*. Nearly one-third of the figures of herbs are new. The rest are copied on a reduced scale from the *German Herbarius*, and the drawing, which is by no means improved, often shows that the copyist did not fully understand the nature of the object which he was attempting to portray. As an example of a delineation which has lost much of its character in copying, we may take the dodder (cf. figs. 89, p. 195; 92, p. 198).

The *Ortus sanitatis* is very rich in pictures. The first edition opens with a full-page woodcut, modified from that at the beginning of the *German Herbarius*, and representing a group

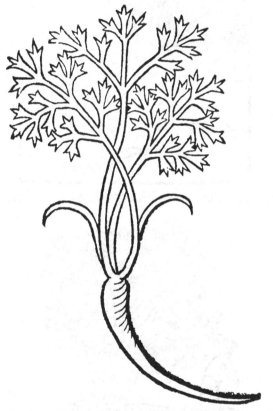

Fig. 9. "Daucus", Carrot [*Ortus sanitatis*, 1491]

of figures, who appear to be engaged in discussing some medical or botanical problem. Before the treatise on animals, there is another large engraving of three persons with a number of beasts at their feet, and, before that on birds, there is a lively picture with an architectural background, showing a scene which swarms with innumerable birds of all kinds,

Fig. 10. "Passer", Sparrow [*Ortus sanitatis*, 1491]

Fig. 11. "Pavo", Peacock [*Ortus sanitatis*, 1491]

whose peculiarities are apparently being discussed by two
savants in the foreground. The treatise on fishes begins with a
landscape with water, enlivened with ships, fishes, crabs, and
mythical creatures, such as mermen. Before the treatise on

Fig. 12. "Bauser vel Bausor" [*Ortus sanitatis,* 1491]

stones, there is a very spirited scene, representing a number
of people in a jeweller's shop, and two large woodcuts of
doctors and their patients illustrate the medical portion with
which the book concludes.

In the treatise upon plants there are various pictures dis-
playing a liveliness of imagination which one misses in

31

modern botanical books. A tree called "Bausor", for instance, which was believed, like the fabulous upas tree, to
exhale a narcotic poison, has two men lying beneath its
branches, apparently in the sleep of death (fig. 12, p. 31). The

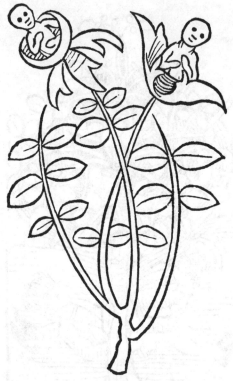

Fig. 13. "Narcissus" [*Ortus sanitatis*, 1491]

engraving which is named "Narcissus" (fig. 13) has dimi-
nutive figures burgeoning from the flowers, as in a transformation scene at a pantomime. It is probably, however,
intended to symbolise the fate of the beautiful youth who
loved his own image too well.

The tree-of-knowledge, with Adam and Eve and the apple,
or with a woman-headed serpent (fig. 14, p. 33), and the
tree-of-life, are considered among other botanical objects.
We are promised, enchantingly, that he who should eat of

the fruit of the tree-of-life "should be clothed with blessed immortality, and should not be fatigued with infirmity, or anxiety, or lassitude, or weariness of trouble".

Among the herbs, substances such as starch, vinegar, cheese, soap, etc., are included, and as these do not lend

Fig. 14. "Arbor...vel lignum scientie",
Tree-of-knowledge [*Ortus sanitatis*, 1491]

themselves to direct representation, they become the excuse for a delightful set of genre pictures. "Wine" is illustrated by a man gazing at a glass; "Bread", by a housewife with loaves on the table before her (fig. 15, p. 34); "Water", by a fountain; "Honey", by a boy who seems to be extracting it from a comb; and "Milk", by a woman milking a cow. The picture which appears under the heading of "Amber" is an example of condensed information (fig. 16, p. 35). The writer points out that this substance, according to some authors, is the fruit or gum of a tree growing by the sea, while

Fig. 15. "Panis", Bread [*Ortus sanitatis*, 1491]

according to others it is produced by a fish or by sea foam. In order to represent all these possibilities, the figure shows the sea, indicated in a conventional fashion, with a tree growing out of it, and a fish swimming in it.

In the treatises on animals and fishes, pictures of mythical creatures are to be found—a fight between a man and hydras, the phoenix in the flames, and a harpy with its claws in a man's body. Other woodcuts show a dragon, the Basilisk, Pegasus, and a bird with a long neck, which is tied in an ornamental knot.

Fig. 16. "Ambra", Amber [*Ortus sanitatis*, 1491]

About the year 1500 a version of the *Hortus* was printed by Antoine Vérard in Paris under the title, *Ortus sanitatis translate de latin en francois*. Henry VII was one of Vérard's patrons, and in the account books of John Heron, Treasurer of the Chamber, which are preserved at the Record Office, there is an entry (1501–2) which runs, "Item to Anthony

Vérard for two bokes called the gardyn of helth...*£6.*"
This refers to a copy, in two parts, of Vérard's translation of
the *Ortus sanitatis,* which is still preserved in the British
Museum.

Fig. 17. "Persica", *Prunus persica* Stokes,
Peach [*Ortus sanitatis,* ? 1497]

The complete *Ortus sanitatis* made its appearance for the
last time as *Le jardin de sante,* printed by Philippe le Noir
about 1539, and sold in Paris, "a lenseigne de la Rose
blanche couronnee". Fig. 18, p. 37, taken from this book,
shows how the artist of the period represented a "Garden of
Health". The title-pages of early herbals were often decor-
ated with such pictures. A more ambitious example is
reproduced in fig. 131, p. 269. In this illustration, an apothe-

cary's storeroom is also portrayed, and a housewife is seen laying fragrant herbs among linen. There is another amusing little garden scene on the title-page of *The grete herball*, published in 1526 (fig. 130, p. 267).

Fig. 18. Wood-cut from the title-page [*Le jardin de sante*, ? 1539]

Chapter III

THE EARLY HISTORY OF THE HERBAL
IN ENGLAND

1. The *HERBARIUM* of Apuleius Platonicus

oncerning the *Herbarium* of Apuleius Platonicus, something has already been said. This treatise was perhaps the first which opened to the English the herbal medicine of southern Europe, and thus connected Britain with the main stream in the history of systematic botany. For this reason it may be considered in this chapter, although manuscript herbals do not, in general, come within our province. In the British Museum, there is an Anglo-Saxon codex[1] of Apuleius with pictures—probably transcribed between A.D. 1000 and the Norman Conquest—which was rendered into modern English in the nineteenth century.[2] We will use this version for quotation, but with the proviso that a new and more accurate translation is greatly to be desired.

It is with the virtues of the herbs that this Anglo-Saxon manuscript is, in the main, concerned; there is little attempt at describing them botanically. In the *Herbarium* of Apuleius, as in other early works, plants were regarded merely as "simples"—that is, as the simple constituents of compound medicines. Jerome Bock, in the middle of the sixteenth century, described his herbal as being an account of "the individual herbs of the earth, called simples" ("die Einfache

[1] Cotton Vitellius C iii.   [2] Cockayne, T. O. (1864): see Appendix II.

38

erd Gewåchs, Simplicia genant"). The term "simple", now
almost obsolete, was a household word at a time when most
remedies were manufactured at home in the stillroom. The
expression of Jaques in *As You Like It*—"a melancholy of
mine own, compounded of many simples, extracted from
many objects"—would not have seemed in the least far-
fetched to an audience of that day. Although the word
"simple", used in this sense, has vanished from our talk, i'.s
antithesis "compound" has held its place in the language of
pharmacy, and, to some extent, in our common speech.

The southern source of the herbal of Apuleius is attested
by the fact that the origin of the healing art is attributed to
Aesculapius and Chiron. We are told, also, that the worm-
woods were discovered by Diana, who "delivered their
powers and leechdom to Chiron, the centaur, who first from
these worts set forth a leechdom". The lily-of-the valley, on
the other hand, is said to have been found by Apollo and given
by him "to Aesculapius, the leech".

Many of the accounts of the virtues of the plants are of the
nature of spells or charms rather than of medical recipes. For
instance it is recommended that "if any propose a journey,
then let him take to him in hand this wort artemisia [worm-
wood],...then he will not feel much toil in his journey".
Even in the twentieth century, wormwood is not unknown as
a charm against the dangers of travel. As recently as 1925, it
was recorded that a driver of the "Auto-Post", on a pre-
cipitous road with hairpin bends leading to Maloja, was
observed to have a branch of this plant hanging from his
wind-screen.

As is usually the case in the older herbals, the proper mode
of uprooting the mandrake is described by Apuleius with
much gusto. "This wort...is mickle and illustrious of aspect,
and it is beneficial. Thou shalt in this manner take it, when
thou comest to it, then thou understandest it by this, that it
shineth at night altogether like a lamp. When first thou seest

its head, then inscribe thou it instantly with iron, lest it fly from thee; its virtue is so mickle and so famous, that it will immediately flee from an unclean man, when he cometh to it; hence as we before said, do thou inscribe it with iron, and so shalt thou delve about it, as that thou touch it not with the iron, but thou shalt earnestly with an ivory staff delve the earth. And when thou seest its hands and its feet, then tie thou it up. Then take the other end and tie it to a dog's neck, so that the hound be hungry; next cast meat before him, so that he may not reach it, except he jerk up the wort with him. Of this wort it is said, that it hath so mickle might, that what thing soever tuggeth it up, that it shall soon in the same manner be deceived. Therefore, as soon as thou see that it be jerked up, and have possession of it, take it immediately in hand, and twist it, and wring the ooze out of its leaves into a glass ampulla."

The writer of the herbal evidently accepted the mythical notion that the mandrake was furnished with human limbs. Pl. v, f. p. 18, shows how this plant was depicted in an early printed edition of the *Herbarium* of Apuleius, but much more spirited and sensational treatments of the same subject are to be found in some of the manuscripts dealing with herbs. Representations from sixteenth-century books are copied in figs. 125, p. 248, and 130, p. 267. The horror of mandrake cannot be dismissed as a superstition of the remote past, for this fear may be encountered in full force to-day. A recent book[1] about Palestine records that a woman of Artas was asked to dig up a mandrake. She agreed to do so, but while she was engaged upon the task, one of her relations passed by and cried, "What are you doing, O woman, digging up that root? Do you not know that there is a little black man there and if you pull him up right down to his feet you will fall ill and take to your bed?" In Britain some of the folk-lore connected with the mandrake has been transferred to the white-bryony.

[1] Crowfoot, G. M. & Baldensperger, L. (1932): see Appendix II.

Another manuscript of Apuleius, of British provenance, is a Latin version, written probably at Bury St Edmunds within fifty years after the Norman Conquest. It is now accessible to students in a facsimile reproduction. It is illustrated, in the main, with figures of the conventional type usually associated with this treatise, but a few show greater realism. One of these—a charming study of a bramble in fruit—has a certain resemblance to the picture of the same plant in the Anicia Juliana Codex of Dioscorides.

2. BANCKES' *HERBALL*

The earliest English printed book containing information of a definitely botanical character, is probably the translation of the *Liber de proprietatibus rerum* of Bartholomaeus Anglicus, which came from Wynkyn de Worde's press before the end of the fifteenth century. A woodcut from it is shown in fig. 19, p. 42. The first book printed in Britain which is a herbal in the strict sense, is, however, an anonymous quarto, without illustrations published in 1525. The title-page runs, *Here begynneth a newe mater, the whiche sheweth and treateth of ẙ vertues and proprytes of herbes, the whiche is called an Herball.* On the last page we find the words "Imprynted by me Rycharde Banckes, dwellynge in London, a lytel fro ẙ Stockes in ẙ Pultry." It is possible that this book may have some claim to originality, but it is more probable that it is derived from an unknown mediaeval manuscript dealing with herbs. It is certainly quite a different work from *The grete herball*, printed in the succeeding year, and, although there are no figures, it is in some ways a better book. Distinctly less space, in proportion, is devoted to the virtues of the plants, and, on the whole, more botanical information is given. For instance, under the heading "Capillus veneris", we find the following description: "This herbe is called Mayden heere or waterworte. This herbe hathe leves lyke to Ferne, but the leves be smaller, and it groweth on walles and stones, and in

III. The Earlier English Herbals

ỹ myddes of ỹ lefe is as it were blacke heere." *The grete herball*, on the other hand, vouchsafes only the meagre information, "Capillus veneris is an herbe so named".

Fig. 19. Woodcut of Herbs and Trees [Bartholomaeus Anglicus, *Liber de proprietatibus rerum*, Wynkyn de Worde, ?1495] *Reduced*

In cases where the virtues of the herbs are not strictly medicinal, they are described in Banckes' *Herball* with more than a touch of poetry. Rosemary has perhaps the most charming list of attributes, some of which are worth quoting. The reader is directed to "take the flowres and make powder therof and bynde it to the ryght arme in a lynen clothe, and it shall make the lyght and mery.. . . Also take the flowres and put them in a chest amonge youre clothes or amonge bokes and moughtes [moths] shall not hurte them.. . . Also boyle the leves in whyte wyne and wasshe thy face therwith. . . thou shall have a fayre face. Also put the leves under thy beddes heed, and thou shalbe delyvered of all evyll dremes.. . . Also take the leves and put them into a vessel of wyne. . . yf thou sell that wyne, thou shall have good lucke and spede in the sale.. . . Also make the a box of the wood and smell to it and it shall preser[v]e thy youthe. Also put therof in thy doores or in thy howse and thou shalbe without daunger of Adders and other venymous serpentes. Also make the a barell therof and drynke thou of the drynke that standeth therin and thou nedes to fere no poyson that shall hurte ỹ, and yf thou set it in thy garden kepe it honestly for it is moche profytable."

The popularity of Banckes' *Herball* is attested by the fact that a large number of editions appeared from different presses, although their identity has been obscured by the various names under which they were published. To consider these editions in detail is a task for the bibliographer rather than for the botanist, and it will not be attempted here. We may, however, mention a few typical examples.

In 1555, or later, a book was printed by "Jhon kynge" with the title: *A litle Herball of the properties of Herbes, newly amended and corrected, wyth certayn Additions at the ende of the boke, declaring what Herbes hath influence of certain Sterres and constellations, wherby maye be chosen the best and most lucky tymes and dayes of their ministracion, according to the Moone beyng in the signes of heaven the which is daily appointed in the Almanacke, made and gathered in the yeare of our Lorde God. MDL[1] the XII daye of February, by Anthony Askham, Physycyon.* This work, which is generally called Askham's *Herball*, is directly derived from Banckes' *Herball*. It is

---

[1] This cannot be the date of printing, as King had not then begun to print; the reference is probably to the date of the *Almanacke*.

rather surprising to discover that, despite the statement on the title-page, the book contains no astrological lore at all.

The book known as Copland's *Herball*, which was probably first published about the same time as Askham's *Herball*, is simply a later edition of the herbal of Rycharde Banckes, and another closely similar edition, with an almost identical title, was published by Kynge.

Another version of the same work, undated, and printed by Robert Wyer, appeared under an even more deceptive title: *A new Herball of Macer, Translated out of Laten in to Englysshe*. This is doubly confusing, because there was not only a certain Aemilius Macer, a contemporary of Virgil and Ovid, who wrote about plants in Latin verse, but there was also a mediaeval herbalist, possibly of the tenth century, whose real name is believed to have been Odo, but who adopted "Macer Floridus", or "Aemilius Macer" as his pseudonym. His work, *De viribus herbarum*, was first printed at Naples in 1477; in this edition the virtues of eighty-eight herbs, spices, etc., are described in Latin verse. There seems to be no justification whatever for the use of the name either of the classical or the mediaeval Macer on the title-page of *A new Herball of Macer*; except for some slight verbal alterations, it is identical with Banckes' *Herball* of 1525. Another closely similar edition, also undated, was published under the name of *Macers Herbal. Practysyd by Doctor Lynacro*. Like that of Macer, the name of Linacre was borrowed, after the fashion of the time, to give the books a well-sounding title, and thus to increase the chances of sale.

3. THE GRETE HERBALL

Among the earlier English herbals, the highest reputation belongs, not to Banckes' *Herball* in any of its forms, but to *The grete herball*. This book was printed by Peter Treveris in 1526 and again in 1529; one of its corrected proof sheets was recently discovered at Queen's College, Oxford, in the binding of an indenture dated 1526, thus affording us a glimpse into the actual making of the book four hundred years ago. *The grete herball* put forward no claim to originality; after the index there is a note, "Thus endeth the grete herball with his tables which is translated out of ẙ Frensshe into Englysshe." The source was, at least in the main, *Le grand Herbier*, to which we have referred on pp. 26, 28. The plant figures in *The grete herball* are degraded copies of the series which first appeared in the *Herbarius zu Teutsch* (e.g. fig. 21, p. 49).

The introduction to *The grete herball*, though it is less naïve and charming than the corresponding part of the *German Herbarius*, may yet be quoted, in part, as giving a definite idea of the utilitarian point of view of the herbalist of the period, and also as bringing home to the reader the far-reaching influence of the theory of the four elements:

"Consyderynge the grete goodnesse of almyghty god creatour of heven and erthe, and al thynge therin comprehended to whom be eternall laude and prays. etc. Consyderynge the cours and nature of the foure elementes and qualytees where to ẙ nature of man is inclyned, out of the whiche elementes issueth dyvers qualytees infyrmytees and dyseases in the corporate body of man, but god of his goodnesse that is creatour of all thynges hath ordeyned for mankynde (whiche he hath created to his owne lykenesse) for the grete and tender love, which he hath unto hym to whom all thinges erthely he hath ordeyned to be obeysant, for the sustentacyon and helthe of his lovynge creature mankynde whiche is onely made egally of the foure elementes and qualitees of the same, and whan any of these foure habounde or hath more domynacyon, the one than the other it constrayneth ẙ body of man to grete infyrmytees or dyseases, for the whiche ye eternall god hath gyven of his haboundante grace, vertues in all maner of herbes to cure and heale all maner of sekenesses or infyrmytes to hym befallyng thrugh the influent course of the foure elementes beforesayd, and of the corrupcyons and ẙ venymous ayres contrarye ẙ helthe of man. Also of onholsam meates or drynkes, or holsam meates or drynkes taken ontemperatly whiche be called surfetes that brengeth a man sone to grete dyseases or sekenesse, whiche dyseases ben of nombre and ompossyble to be rehersed, and fortune as well in vilages where as nother surgeons nor phisicians be dwellyng nygh by many a myle, as it dooth in good townes where thev

Fig. 20. "Yvery", Ivory [*The grete herball*, 1529]

be redy at hande. Wherfore brotherly love compelleth me to wryte thrugh ẙ gyftes of the holy gost shewynge and enformynge how man may be holpen with grene herbes of the gardyn and wedys of ẙ feldys as well as by costly receptes of the potycarys prepayred."

The conclusion of the whole matter, which is set forth immediately before the index, is in these words:

"O ye worthy reders or practicyens to whome this noble volume is present I beseche yow take intellygence and beholde ẙ workes and operacyons of almyghty god which

The grete herball

hath endewed his symple creature mankynde with the graces
of ỹ holy goost to have parfyte knowlege and understandynge
of the vertue of all maner of herbes and trees in this booke
comprehendyd."[1]

From the twentieth-century point of view, *The grete
herball* contains much that is curious, especially in relation
to medical matters. Bathing was evidently regarded as a
strange fad. We learn, on the authority of Galen, that "many
folke that hath bathed them in colde water have dyed or they
came home". Water drinking seems to have been thought
almost equally pernicious, for we are told, "Mayster Isaac
sayth that it is unpossyble for them that drynketh overmoche
water in theyr youth to come to the aege that god ordeyned
them." A period when men were more prone than they are
to-day to settle their differences with the aid of their own
fists, is reflected in the various remedies for "blackenesse or
brusinge comynge of strypes, specyally yf they be in the
face". *The grete herball* is thoroughly mediaeval in character,
despite its date of publication, and the number of prescriptions
against melancholy found in its pages is probably an indication of the miserable conditions of life in the middle ages.
"To make folke mery at ye table", we are told to "take foure
leves and foure rotes of vervayne in wyne, than spryncle the
wyne all about ỹ hous where the eatynge is and they shall be
all mery." The smoke of "Aristologia", "maketh the pacyent
mery mervaylously", and also "dryveth all devyllsshnesse
and all trouble out of the house". Bugloss and mugwort are
also recommended to produce merriment, and it is suggested
that the lesser-mugwort should be laid under the door of the
house, for, if this is done, "man nor woman can not anoy in
that hous".[2] The reader is told how to make a "ptysan" of
barley-water; how to gargle the throat; and how to dis-

---

[1] These quotations are from the edition of 1526, but all subsequent quotations from
that of 1529.

[2] The expression in the French original is, "homme ne femme ne pourra nuire en
ceste maison".

tinguish good from inferior musk. Moreover he is given recipes against such ills as forgetfulness and fear. Hair-dyes and stains for the nails also receive their share of attention.

Remarkable powers were attributed to products of the ocean, such as coral and pearls. The former is described as being "a maner of stony substaunce that is founde in partyes of the see, and specyally in holowe, and cavy hylles that ben in the see, and groweth as a maner of a glewy humoure, and cleveth to the stones". The writer mentions that "some say that the reed corall kepeth the hous that it is in fro lyghtenyng, thondre, and tempest". Pearls were regarded as of great value in medicine; for weakness of the heart, the patient is recommended to "Take the powdre of perles with sugre of roses"—a remedy which should please the most fastidious. Many travellers' tales are incorporated in the herbal; we find, for instance, a thrilling description of the lodestone. "Lapis magnetis is the adamant stone that draweth yren. It...is founde in the brymmes of the occyan see. And there be hylles of it, and these hylles drawe ẏ shyppes that have nayles of yren to them, and breke the shyppes by drawynge of the nayles out." This description is illustrated by a picture of a rocky pinnacle and a ship going to pieces; one man is already in the water, and two others are on the point of losing their lives.

The remedies for various ailments strike the modern reader as being violent in a terrifying degree, and adapted to a more robust age than the present. They incline one to echo the words, "There were giants in the earth in those days"; but apparently the sixteenth century held an exactly corresponding view of its predecessors, for under the heading of "whyte elebore" we read, "In olde tyme it was commely used in medycyns as we use squamony. For the body of man was stronger than it is now, and myght better endure the vyolence of elebore, for man is weyker at this time of nature."

The grete herball

In *The grete herball* Greek mythology finds a place, side
by side with Christianity. The discovery of wormwood
(*Artemisia*) is attributed, as in the herbal of Apuleius, to
Diana, who gave the plant to the centaurs; but, in the event
of being bitten by a mad dog, the sufferer is recommended to
appeal to the Virgin Mary before employing any remedy:
"As sone as ye be byten go to the chyrche, and make thy

De Nenufare.

Fig. 21. "Nenufar", Water-lily [*The grete herball, 1529*]

offrynge to our lady, and pray here to helpe and hele the.
Than rubbe ye sore with a newe clothe," etc.

Quite a number of medicines enumerated in *The grete
herball* have remained in use until our own times. Liquorice
and horehound are recommended for coughs; laudanum,
henbane, opium and lettuces as narcotics; olive oil and slaked
lime for scalds; cuttle-fish bone for whitening the teeth; and
borax and rose water for the complexion.

This book throws an interesting light on the early names of
British plants. The primrose is called "Prymerolles" or

"saynt peterwort". The "devylles bytte" is said to be "so called bycause the rote is blacke and semeth that it is iagged with bytynge, and some say that the devyll had envy at the vertue therof and bete the rote so for to have destroyed it". Duckweed is called "Lentylles of the water" or "frogges fote", while wild-arum is known by the descriptive title of "prestes hode", and wood-sorrel is named "Alleluya" or "cuckowes meate".

One of the most noticeable features of the herbal is the exposure of methods of "faking" drugs, for the protection of the public, "to eschew ẙ fraude of them that selleth it". This is a great step in advance from the days of the old Greek herbalists, when secrecy was part of the stock-in-trade of a druggist, and, as we have pointed out in a previous chapter, the credulous public was warned off by threats of the miraculous and fearful ills which would follow any unskilled meddling with the subject.

Another work illustrated with the same figures as those of *The grete herball,* was *The vertuose boke of Distyllacyon of the waters of all maner of Herbes,* issued in 1527. This was an English version of the *Liber de arte distillandi de simplicibus* of Jerome of Brunswick (Hieronymus Braunschweig). In the prologue the publisher, Laurence Andrew, tells us that he undertook to make this translation "beynge moved with naturall love unto my contre, whiche surely shold want yf I were able to performe it, no profytable booke for lacke of a Translatour, that is in any other language wrytten"; and he adds, "Spare nat favorable reder to peruse and revolve to thy synguler helthe, conforte, and lernynge, thys booke of distyllacion. Lerne the hyghe and mervelous vertu of herbes, knowe how inestymable a preservatyve to the helth of man god hath provyded growyng every day at our hande, use the effects with reverence, and gyve thankes to thy maker celestyall." As might have been expected from this introduction, the treatise is almost entirely occupied with

The vertuose boke of Distyllacyon

methods of distillation, and medical directions, but occasionally there is an effective touch of description; the mistletoe
leaf, for instance, is said to be "nother ful greŕe, nor full
yelowe". The colophon tells us that the book was printed "in
the flete strete...in the sygne of the golden Crosse...
Goddys grace shall ever endure".

Chapter IV

THE BOTANICAL RENAISSANCE OF THE SIXTEENTH AND SEVENTEENTH CENTURIES

1. THE HERBAL IN GERMANY

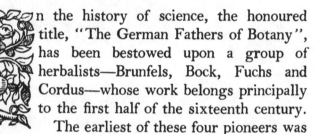In the history of science, the honoured title, "The German Fathers of Botany", has been bestowed upon a group of herbalists—Brunfels, Bock, Fuchs and Cordus—whose work belongs principally to the first half of the sixteenth century.

The earliest of these four pioneers was Otto Brunfels [Otho Brunfelsius] (pl. vi), who owed his surname to the fact that the family to which his father, a cooper, belonged, came from Braunfels near Mainz. Dates as far apart as 1464 and 1490 have been given for his birth by different authorities. When he grew up, he entered a Carthusian monastery, but after some years of this life he became converted to Lutheran principles. He fled from the cloister in 1521, and, after various wanderings and intermittent preaching, he took up his abode in Strasburg, where he was for nine years a school teacher. He wrote a number of theological works, but ultimately turned his attention to medicine, and, shortly before his death, which took place in 1534, he had become town physician at Bern.

A new era in the history of the herbal may be said to date from the year 1530, when the first part of the *Herbarum vivae eicones* was published by Schott of Strasburg. This work is

Plate vi

OTTO BRUNFELS (d. 1534), aged 46
[Engraving in the Hortus Bergianus, Stockholm]

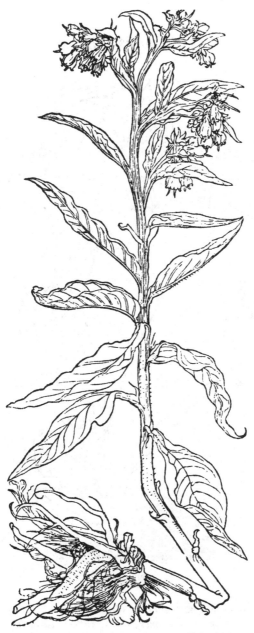

Fig. 22. "Walwurtz männlin", *Symphytum officinale* L., var. *purpureum*
Pers., Comfrey [Brunfels, *Herbarum vivae eicones*, vol. I, 1530] *Reduced*

Helleborus Niger.

Chriſtwurtz.

Fig. 23. "Helleborus Niger", *H. viridis* L., Green-hellebore [Brunfels,
Herbarum vivae eicones, vol. i, 1530] *Reduced*

commonly called Brunfels' herbal, but it would be juster to associate it with the name of the artist, Hans Weiditz, who was responsible for the illustrations, in which there was, as the title indicates, a real return to nature. The plants are represented as they are, and not in the conventionalised aspect which had become traditional in the earlier herbals through successive copying of each drawing from a previous one, without reference to the plants themselves. Examples of the woodcuts in the *Herbarum vivae eicones* are shown in figs. 22, p. 53; 23, p. 54; 24, p. 56; 25, p. 57; 97, p. 205; 98, p. 207; 99, p. 208. In a later chapter (p. 206) we will consider a recent discovery which has thrown light on the work of Weiditz.

The pictures in the *Herbarum vivae eicones* are incomparably better than the text. Brunfels' knowledge of botany was chiefly derived from the study of certain Italian authors—Manardus, and others—who had occupied themselves in identifying the plants of their own peninsula with those described by Dioscorides. When Brunfels attempted to employ the same methods in his examination of the flora of the Strasburg district, and of the left bank of the Rhine, many difficulties and discrepancies inevitably arose. He had no understanding of the geographical distribution of plants, and he failed to realise that different regions have dissimilar floras. This failure is the more surprising, when we remember that Theophrastus, eighteen hundred years earlier, had pointed out that the provinces of Asia have each their own characteristic plants, and that some, which occur in one region, are absent from another. The narrowness of Brunfels' outlook is shown also by the fact that he thinks it necessary to apologise for including the lovely woodcut of the pasque-flower (reproduced in fig. 98, p. 207), because this herb is not used by apothecaries, and has no Latin name. The unfortunate plants which are in this predicament, he dismisses as "herbae nudae".

Jerome Bock, who in his Latin writings called himself

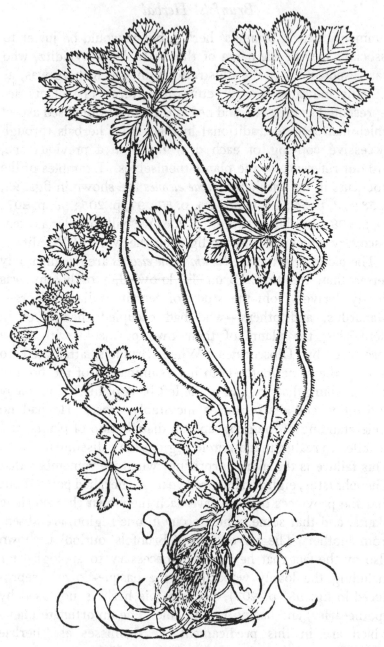

Fig. 24. "Synnaw", *Alchemilla vulgaris* var. β L., Ladies'-mantle
[Brunfels, *Herbarum vivae eicones*, vol. II, 1531] *Reduced*

Benedicten wurtzel.

Fig. 25. "Caryophyllata", *Geum urbanum* L., Avens [Brunfels,
Herbarum vivae eicones, vol. III, 1540] *Reduced*

Hieronymus Tragus (fig. 26), was a contemporary of Brunfels, though his botanical work was somewhat later in date.

Fig. 26. Hieronymus Bock or Tragus, 1498–1554 [Engraving by David Kandel. *Kreuter Bůch*, 1551]

He was born in 1498, and destined by his parents for the cloister; but he proved to have no vocation for the monastic life, and, having passed through a university course, he obtained, by favour of the Count Palatine Ludwig, the post of school teacher at Zweibrücken, and overseer of the Count's garden. After his patron's death, he removed to Hornbach, where he was Lutheran pastor, and also practised medicine, while devoting his spare time to botany. Eventually, however, the counter-reformation drove him from Hornbach. He was in serious straits until Count Philip of Nassau-Saarbrücken, whom he had previously cured of a severe illness, gave him shelter and support in his own castle. Finally he was able to return to Hornbach, where he filled the office of preacher until his death in 1554.

Bock's great work is the *New Kreütter Bůch*, which appeared in 1539, printed at Strasburg by Wendel Rihel. The first edition was without illustrations, but others, issued in 1546 and later, contained many woodcuts (e.g. figs. 27, p. 60; 28, p. 62; 29, p. 63; 106, p. 220; 107, p. 221; 108, p. 222). Although some of these cuts are founded on the figures of Brunfels and Fuchs, others were specially drawn and engraved by David Kandel, whose initials are seen on the portrait of Bock reproduced in fig. 26, p. 58.

Bock's chief claim to remembrance does not, however, depend upon his illustrations, but upon his admirable descriptions, written in the plain, racy German of the people. He noted, moreover, the mode of occurrence and localities of the plants to which he referred, and in this feature his work showed some approach to a flora in the modern sense. He looked at the world around him with his own eyes, rather than with those of bygone classical authorities, and the excellence of his observations is often surprising. He detected, for instance, that the two-headed eagle, which is revealed when a fern rhizome is cut across, is nothing more than the system of veins seen in section. Bock's sturdy independence of mind made him reject superstitions, such as the "monkey tricks and ceremonies" connected with the use of wormwood, and it also led him to make a serious attempt to get at the truth about fern seed. Dioscorides and later writers

Von Erdberen. cap. clxx.

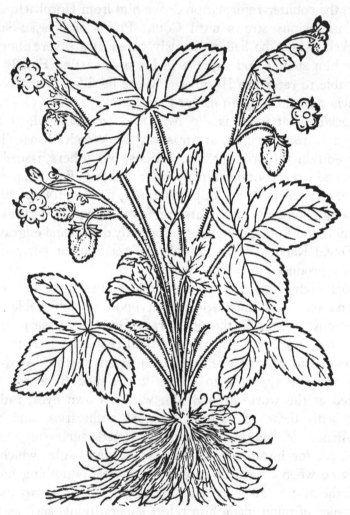

Fig. 27. "Erdberen", *Fragaria*, Strawberry [Bock, *Kreuter Buch*, 1546]

had stated that ferns possessed no fruit or seed; it was commonly held, however, by country people in Bock's day, that fern seed was produced at midnight on midsummer eve, but that to gather it was attended with danger, unless certain superstitious rites were observed. Bock's account of his experiment may be quoted in the words of a sixteenth-century translation: "I have foure yeres together one after an other upon the vigill of saynt John the Baptiste (whiche we call in Englishe mydsomer even) soughte for this sede of Brakes upon the nyghte, and in dede I fownde it earlye in the mornynge before the daye brake, the sede was small blacke and lyke unto poppye. I gatherid it after this maner: I laide shetes and mollen leaves underneth the brakes whiche receyved the sede.... I went aboute this busynes, all figures coniurynges, saunters, charmes, wytchcrafte, and sorseryes sett a syde, takynge wyth me two or three honest men to bere me companye." Having been at so much pains to arrive at the fact that Dioscorides was wrong, and the country people were right, about the existence of fern seed, it is a pity that Bock did not go just one step further, and try to discover whether seed could be obtained at times other than midsummer eve; but this possibility did not, apparently, enter his head.

Bock's scientific caution and instinct for experiment comes out clearly in his remarks about the seed of willows. He says that he does not know whether the "wool of willows" *always* takes the place of seed or not, but he knows this to be true of *one* kind, since he has succeeded in growing willows of the same kind from the "wool".

Otto Brunfels deserves especial gratitude for the way in which he encouraged Bock to write his herbal. Late in his life, Brunfels trudged on foot the forty miles from Strasburg to Hornbach, to see Bock's gardens and collections; and from that time ceased not to urge him to overcome his modest scruples at the greatness of the task, and to reduce his materials to writing for the service of the fatherland.

Pimpernuß.

Fig. 28. "Pimpernuss", *Staphylea pinnata* L.,
Bladder-nut [Bock, *Kreuter Buch*, 1546]

De Tribulo aquatico.

Waſſernuß.

Fig. 29. "Tribulus aquaticus", *Trapa natans* L.,
Bull-nut [Bock, *De stirpium*, 1552]

63

IV. The Botanical Renaissance

Leonhart Fuchs [Fuchsius], the third of the German Fathers of Botany (fig. 30, p. 65), belonged to the same generation as Jerome Bock, though he was a little younger, and produced his chief botanical work three years later. He was born in 1501 at Wemding in Bavaria, and, while still almost a child, became a student at the University of Erfurt, matriculating, it is said, in his thirteenth year. After a period as a teacher he resumed his studies, this time at the University of Ingolstadt, where he devoted himself chiefly to classics, and became a Master of Arts. After this he turned his attention to medicine, and took a doctor's degree. At Ingolstadt he came under the influence of Luther's writings, which won him over to the reformed faith.

Fuchs began to practise as a physician at Munich, but in 1526 he returned to Ingolstadt as Professor of Medicine. In 1535 he was appointed to a chair at Tübingen, and, while he held this post, he declined a call to the University of Pisa, and also an invitation to become physician to the King of Denmark. It is clear that, both as practitioner and teacher, he was in great demand. He acquired a widespread reputation by his successful treatment of a terrible epidemic disease, which swept over Germany in 1529. A little book of medical instructions and prayers against the plague, which was published in London in the latter half of the sixteenth century, shows that his fame had extended to England. It is entitled, *A worthy practise of the moste learned Phisition Maister Leonerd Fuchsius, Doctor in Phisicke, most necessary in this needfull tyme of our visitation, for the comforte of all good and faythfull people, both olde and yonge, both for the sicke and for them that woulde avoyde the daunger of the contagion.*

In spite of his professional activity, Fuchs found time to produce a botanical masterpiece, which appeared in 1542 from the press of Isingrin of Basle, under the title *De historia stirpium*. This was a Latin herbal dealing with about four hundred native German, and one hundred foreign plants; it

64

Fig. 30. Leonhart Fuchs (1501–66)
[*De historia stirpium*, 1542] *Reduced*

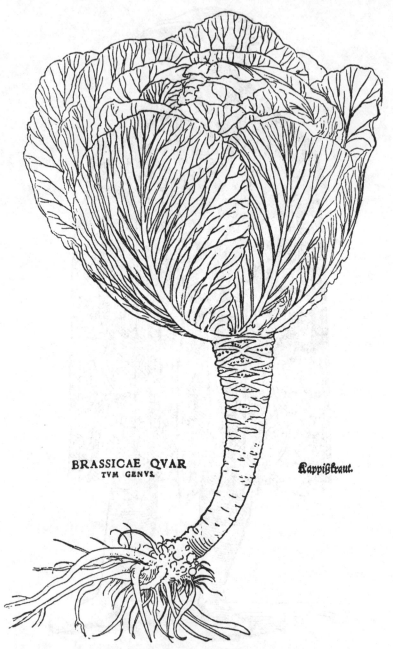

BRASSICAE QVAR
TVM GENVS.

Rappißkraut.

Fig. 31. "Brassicae quartum genus", Cabbage [Fuchs, *De historia stirpium*, 1542] *Reduced*

66

was followed in the succeeding year by a German edition, called the *New Kreüterbüch*. Fuchs is notably superior to his two predecessors in matters calling for scholarship, such as the critical study of the plant nomenclature of classical authors. His herbal rivals, or even surpasses that of Brunfels in its illustrations, and greatly exceeds it in the number of species described; but it must be remembered that Fuchs was able to make use of the works of Bock and of the Swiss naturalist, Gesner, which were published while his herbal was still in the making.

The preface with which *De historia stirpium* opens, is not only interesting in substance, but is written in singularly pure and fine Latin. Fuchs expresses keen indignation at the ignorance of herbs displayed even by medical men. His outburst on this subject may be translated literally as follows: "But, by Immortal God, is it to be wondered at that kings and princes do not at all regard the pursuit of the investigation of plants, when even the physicians of our time so shrink from it that it is scarcely possible to find one among a hundred who has an accurate knowledge of even so many as a few plants?"

That Fuchs' work was indeed a labour of love is a conviction that must force itself upon everyone who studies his herbal, and it is further borne out by his own words in the preface—words which bear the stamp of a lively enthusiasm: "But there is no reason why I should dilate at greater length upon the pleasantness and delight of acquiring knowledge of plants, since there is no one who does not know that there is nothing in this life pleasanter and more delightful than to wander over woods, mountains, plains, garlanded and adorned with flowerlets and plants of various sorts, and most elegant to boot, and to gaze intently upon them. But it increases that pleasure and delight not a little, if there be added an acquaintance with the virtues and powers of these same plants."

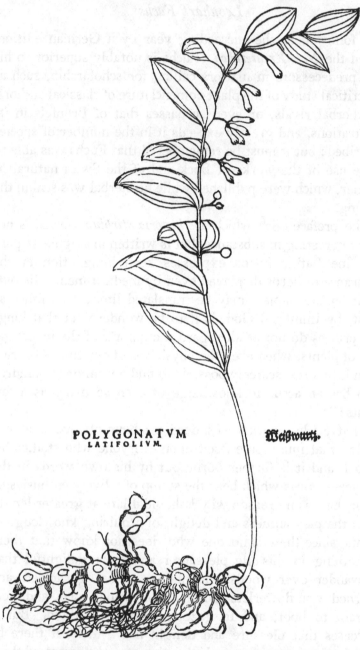

POLYGONATVM
LATIFOLIVM.

Weißwurtz.

Fig. 32. "Polygonatum latifolium", *P. multiflorum* All., Solomon's-seal
[Fuchs, *De historia stirpium*, 1542] *Reduced*

68

Fig. 33. "Cucumis turcicus", *Cucurbita Pepo* L., var. *oblonga* DC.,
Pumpkin-gourd [Fuchs, *De historia stirpium*, 1542] *Reduced*

IV. The Botanical Renaissance

Fuchs arranged his work alphabetically, making no attempt at a natural grouping of the plants, and his herbal is thus without importance in the history of plant classification. The woodcuts which illustrate Fuchs' herbal are of first-rate quality (figs. 31, p. 66; 32, p. 68; 33, p. 69; 65, p. 154; 78, p. 175; 79, p. 178; 80, p. 180; 101, p. 211; 102, p. 213; 103, p. 214; 104, p. 216). Some of them gain a special interest as being the earliest European figures of certain American plants, e.g. the pumpkin-gourd (fig. 33, p. 69) and Indian-corn. The influence of Fuchs' illustrations is more strongly felt in later work than that of his text. The majority of the wood-engravings in Dodoens' *Crüÿdeboeck* (1554), Turner's *New Herball* (1551–68), Lyte's *Nievve Herball* (1578), Jean Bauhin's *Historia plantarum universalis* (1650, 1), and Schinz's *Anleitung*, 1774, are copied from Fuchs, or even printed from his actual wood-blocks, while use was made of his figures in the herbals of Bock, Egenolph, d'Aléchamps, Tabernae-montanus, Gerard, Nylandt, etc., and in the commentaries on Dioscorides of Amatus Lusitanus and Ruellius. It was not the large woodcuts in *De historia stirpium* (1542) which chiefly served for these borrowings, but the smaller versions of the blocks, made for Fuchs' octavo herbal of 1545.

The publisher Christian Egenolph of Frankfort, though not himself a botanical writer, must be mentioned at this stage, because his name is associated with various illustrated books, which had a remarkably wide circulation. His engravings, which were mostly pirated from Brunfels and others, were not even used in a new herbal, but were attached, in the first place, to an edition of the old *German Herbarius*, enlarged and improved by Dr Eucharius Rösslin [Rhodion]. The book was issued under the name of *Kreutterbůch von allem Erdtgewåchs*; its title-page is shown in fig. 34, p. 71.

Egenolph was evidently a keen man of business, with a flair for attractive picture-books which would appeal to the public. In Rösslin's *Kreutterbůch*, not only plants, but all sorts

70

Fig. 34. Title-page [Rösslin (Rhodion), *Kreutterbüch*, 1533] *Reduced*

71

of miscellaneous objects are included; coral, for instance, is illustrated by a drawing of a necklace, and there are a number of amusing little pictures of birds and beasts. Many later editions with similar or related texts, or no letterpress at all, were issued by this publishing house; they were edited, after Rösslin's death, by Theodor Dorsten, and then by Adam Lonitzer [Lonicerus], Egenolph's son-in-law. The title-page of Dorsten's Latin herbal is reproduced in fig. 36, p. 73. No

Erdöpffel.

Vulgago. Panisporcinus. Ciclamen.
Malum terræ. Archanita. Bothormarien.
Schweinbrodt.

Fig. 35. "Erdöpffel", *Ranunculus Ficaria* L., Lesser-celandine
[Rösslin (Rhodion), *Kreutterbûch*, 1533]

other botanical books of the period had as great a success as this series, which, for more than a hundred years was issued by Frankfort presses. Even so long an existence did not exhaust the vitality of these works, and an edition called *Adam Lonicers Krâuter=Buch* appeared at Augsburg in 1783— that is to say, exactly two hundred and fifty years after the *editio princeps* of the sequence.

Egenolph's success was, however, achieved in the teeth of much adverse contemporary criticism. Fuchs, in the preface

BOTANICON,

CONTINENS HERBARVM, ALIORVMQVE

Simplicium, quorum usus in Medicinis est, descriptiones, & Ico-
nas ad uiuum effigiatas : ex præcipuis tam Grecis quàm Latinis
Authoribus iam recens concinnatum. Additis etiam, quæ
Neotericorum obseruationes & experientiæ uel comprobarunt
denuo, uel nuper inuenerunt.

AVT. THEODERICO DOR-
stenio Medico.

Cum Gratia & Priuilegio Cæsareo.

FRANCOFORTI, Christianus Egenolphus
excudebat.

Fig. 36. Title-page [Dorsten, *Botanicon*, 1540] *Reduced*

73

to his *De historia stirpium* (1542), referred with unsparing touch to the botanical mistakes in the Frankfort books. His trenchant indictment may be rendered into English as follows: "Among all the herbals which exist today, there are none which have more of the crassest errors than those which Egenolph, the printer, has already published again and again." This statement Fuchs supports with actual examples.

It must nevertheless be admitted that, even if their quality left much to be desired, the herbals published by Egenolph and his successors did good service in disseminating some knowledge of the plant world among a very wide public, even outside Germany. There is, in the British Museum, a fine example of the 1536 edition, with a binding stamped in gold, and bearing the arms of Mary, Duchess of Suffolk, daughter of Henry VII. The duchess may perhaps have inherited a taste for herbals from her father, for, as we have already mentioned (p. 35), he is known to have bought a copy of Vérard's translation of the *Ortus sanitatis*.

Among the German Fathers of Botany, a fourth name is included, which is comparatively little known—that of Valerius Cordus (1515–44). He was a naturalist of unusual capacity, who, but for his early death, might have become one of the most renowned of the herbalists of the sixteenth century. His father, Euricius Cordus, was a physician, botanist, and man of letters—so Valerius was brought up in an intellectual atmosphere. At sixteen he obtained his bachelor's degree at the University of Marburg, and after studying in various towns, he passed from the position of pupil to that of teacher, and expounded Dioscorides at the University of Wittenberg. He travelled widely in search of plants, and visited many of the savants of the period. He is known to have made a stay at Tübingen, and it can scarcely be doubted that he became acquainted personally with Leonhart Fuchs.

Valerius Cordus had always longed to see, under their

native skies, the plants about which the ancients had written, and, in fulfilment of this dream, he spent two years in Italy. He visited Padua, Bologna, Florence, Siena, and other towns, and late in the summer of 1544 he turned towards Rome, travelling with two or three companions. The journey was very trying to men accustomed to a more northerly climate, and they were attacked by malaria. Wild and hazardous country was traversed in extreme heat, and the exposure and fatigue led to a tragic conclusion. After a kick from a horse, Cordus was seized with fever; his companions had great difficulty in conveying him to Rome, where he died late in September. It was written of him—

mens ipsa recepta est
Coelo; quod terrae est, maxima Roma tenet.

None of the botanical work of Valerius Cordus was published during his lifetime, but, after his death, Gesner edited the *Historia stirpium*, which Cordus had left in manuscript, as well as a commentary on Dioscorides, compiled from the notes of a student who attended Cordus' lectures. The *Historia*, which is a work of outstanding importance, is especially distinguished in the excellence of its descriptions. The author seems to have examined the plants as living things, and for their own sakes—not merely as simples to be used in the arts of healing. Cordus did noteworthy service to medicine, however, for, when he passed through Nuremberg on his travels, he was able to lay before the physicians of that town a collection of medical recipes, chiefly selected from earlier writings. This work, which had for some time been in use in Saxony in manuscript form, was considered so valuable that, after it had been examined and tested under the auspices of the town council, it was published officially at Nuremberg in 1546 as *Pharmacorum...Dispensatorium*.[1] It is said to be

[1] The writer is indebted to Mr T. E. Wallis of the Pharmaceutical Society for showing her the beautiful facsimile version of this rare work, with an introductory account of the various editions, which was published in 1934; see Winckler, L., Appendix II.

the first work of the nature of a pharmacopoeia ever issued under government authority.

A passing reference may be made at this point to Jacob Dietrich [Theodorus] of Bergzabern (?1520–90), a herbalist who is better known as Tabernaemontanus. He was closely connected with the German Fathers of Botany, having been the pupil of Otto Brunfels in boyhood, and afterwards, of Jerome Bock, to whom he refers as "mein lieber *Praeceptor*". Like his two masters, he was a Protestant. Medicine was his profession, and he combined it with the study of botany. He projected a herbal, and spent what he could upon it, but the total cost was beyond his means, and the eventual printing of the book, after he had worked upon it for thirty-six years, was due, in part, to the generosity of the Count Palatine Frederick III, and of the Frankfort publisher, Nicolaus Bassaeus. The herbal appeared in 1588,91, under the title *Neuw Kreuter-buch*, and in 1590 the illustrations were issued without any text as the *Eicones plantarum*. The *Kreuterbuch* is a massive and fully illustrated work, with some attractive woodcut decorations; the title-page states that 3000 plants are described. Its lasting popularity is attested by the fact that a version appeared at as late a date as 1731. The figures are for the most part not original, but are reproduced from Bock, Fuchs, Mattioli, Dodoens, de l'Écluse and de l'Obel. The wood-blocks collected by Tabernaemontanus became familiar in England a few years later, when they were acquired by the printer, John Norton, and used in 1597 to illustrate the first edition of Gerard's *Herball*.

There is yet another German herbalist of the sixteenth century whose work must not be overlooked. This is Joachim Camerarius the Younger (pl. vii). His father, who was a celebrated philologist and the friend and biographer of Melanchthon, adopted the name Kammermeister, or Camerarius, in place of the family name of Liebhard. Joachim the Younger, who was born in 1534, was attracted to botany

Plate vii

JOACHIM CAMERARIUS, the Younger (1534–1598)

[Engraving by Bartholomaeus Kilian, probably between 1650 and 1700; Department of Prints and Drawings, British Museum]

in his early youth. He studied at Wittenberg and other universities, and travelled in Hungary and Italy. He spent some time in the latter country, and took a doctor's degree in

Fig. 37. "Ocimoides fruticosum", *Silene fruticosa* L.
[Camerarius, *Hortus medicus*, 1588]

medicine at Bologna. At Pisa, he became acquainted with Andrea Cesalpino. Finally he returned to Germany, and settled in Nuremberg, his birth-place. Here he cultivated a garden which was kept supplied with rare plants by the merchants of that city, and correspondents in other lands.

IV. The Botanical Renaissance

Camerarius brought out editions of Mattioli (*Kreuterbuch* and *De plantis Epitome*), but his chief work was the *Hortus medicus et philosophicus*, which appeared in 1588. Figs. 37,

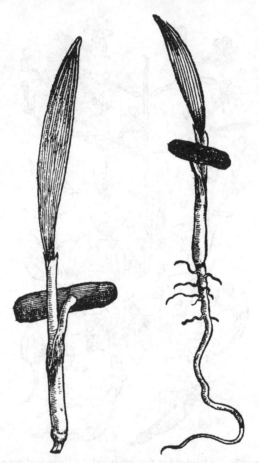

Fig. 38. "Palma", seedlings of *Phoenix dactylifera* L., Date-palm [Camerarius, *Hortus medicus*, 1588]

p. 77; 38, p. 78; 81, p. 182; 119, p. 233, are reproduced from this book. Camerarius was a good observer, and his travels furnished him with information regarding the localities for the plants which he described.

Christophe Plantin

In the sixteenth century, the herbal flourished exceedingly in the Low Countries. This was due not only to the zeal and activity of the botanists of the Netherlands, but also to the munificence, and love of learning for its own sake, which distinguished that prince of publishers, Christophe Plantin of Antwerp.

Plantin's life extended from 1514 to 1588, thus including the years in which the "herbal period" came to full fruition. He was a native of Touraine, and he learned the arts of printing and bookbinding at Caen. Towards 1550, he and his wife, Jeanne Rivière, settled in Antwerp, where he earned his living by bookbinding and other leather work, until one of the commissions with which he was entrusted accidently deflected the whole course of his life. The Secretary to Philip II, having to transmit a precious jewel to the Queen of Spain, asked Plantin to make a leather casket for its conveyance. In the evening of the day on which it was finished, Plantin set forth to deliver it to the Secretary, who required it urgently. On the way he was so ill-starred as to meet a party of masked and drunken revellers, searching for a cithare player against whom they had some grudge. They mistook the package which Plantin was carrying for a musical instrument, and one of them ran him through with his sword, injuring him severely. Though he recovered, he had to face the fact that he would never again possess the strength to do the heavier work of bookbinding; so he resolved to turn to printing, in which he was also expert. Despite crises and vicissitudes which would have driven a lesser man to despair, he succeeded—*Labore et constantia,* as his motto runs—in creating a business which achieved an unique position in the history of printing and publishing. In 1576 he established his presses in a building at one side of the Marché du Vendredi, and in this building printing continued to be carried on until after the middle of the nineteenth century. Plantin's reputation may be gauged by the fact that he received flattering offers, which he refused, both from the Duke of Savoy and from the King of France, to transfer his printing works from the city of his adoption to their territories.

The secret of Plantin's achievement is not easy to analyse. It perhaps lay partly in his intuitive judgment of men, and his capacity for friendship, which enabled him to collect an unrivalled staff. There was also an effective element of ruthlessness in his personality, if we may judge from a passage in one of his own letters, dated December, 1570. He relates that his four elder daughters, "depuis l'aage de quatre à cinq ans jusque à l'aage de douze ans,...nous ont aidé à lire les espreuves de l'imprimerie en quelque escriture et langue qui se soit offerte pour imprimer". It is not surprising that the eldest child, who, after this severe apprenticeship had been sent to Paris, in order that she might learn an elegant script from the royal writing master, was brought home again at the age of twelve, because her sight had failed.

One of Plantin's daughters married Jean Moretus, her father's chief assistant and successor, and from him the business descended through eight generations of printers to Édouard Moretus, the last of the firm, from whom the City of Antwerp purchased the Maison Plantin in 1876—exactly three hundred years after Christophe Plantin first set up his presses there. The house, which had remained essentially unchanged since the days of the founder and his earlier successors, is now preserved as the Musée Plantin-Moretus. It is built round a rectangular courtyard, with such beauty, both in proportion and in detail, that one feels at once that Plantin achieved the ambition which he expressed in his charming sonnet—*Le Bonheur de ce Monde*—"Avoir une maison commode, propre et belle."

The pictures, furniture and hangings, and not only the very presses, fonts, and furnaces for casting the type, but even the old account books and corrected proof-sheets are still to be seen, all in their appropriate places. A visit to the Maison

Fig. 39. Rembert Dodoens, 1517–85 [*A Nievve Herball*, translated by Lyte, 1578]

Plantin is, indeed, an infallible recipe for transporting the imagination back to the time of the later renaissance, when printing, having passed through its earlier struggles, had come to the stage when it was accorded the reverence due to one of the fine arts.

Rembert Dodoens [Dodonaeus] (fig. 39, p. 81), the first Belgian botanist of world-wide renown, was a contemporary of Plantin, having been born at Malines in 1517. He studied at Louvain, and visited the universities and medical schools of France, Italy, and Germany, eventually qualifying as a doctor. In 1574, at the invitation of the Emperor Maximilian II, he became court physician at Vienna. The fact that his friend, Charles de l'Écluse, was already employed there, probably influenced his acceptance of the post. He continued in Vienna as physician to Maximilian's successor, Rudolf II. Then, after a period in Cologne and Antwerp, he was invited in 1582 to a professorship of medicine at Leyden. He died in that town three years later.

Dodoens' interest in the medical aspect of botany led him to write a herbal, and, in order to illustrate it, he obtained the use of the wood-blocks which had been employed in the octavo edition of Fuchs' work. To these a number of new engravings were added. The book was published in Flemish in the year 1554 by Van der Loe, under the title *Crüydeboeck*. The text is not a translation of Fuchs, as is sometimes assumed, although Dodoens was certainly to some extent indebted to the German herbalist. Almost simultaneously with the first Flemish edition, a French issue appeared under the title of *Histoire des plantes*. The translation was carried out by Charles de l'Écluse, with whose own work we shall deal shortly. Dodoens supervised the production of this book, and took the opportunity to make some additions. It became known in this country through Lyte's English version, which will be discussed in a later section of the present chapter.

The last Flemish edition of the herbal for which the author himself was responsible was printed by Van der Loe in 1563. It was Christophe Plantin who published all Dodoens' later books, including his collected works—*Stirpium historiae pemptades sex* (1583). In the *Pemptades*, the botanist in Dodoens was more to the fore, and the physician less in **Capparis.**

Fig. 40. "Capparis", *C. spinosa* L., Caper [Dodoens, *Pemptades*, 1583]

evidence than in his earlier work. It is particularly difficult to appraise with any exactness the services which Dodoens rendered to botany. He and his two younger countrymen, de l'Écluse and de l'Obel, freely imparted their observations to one another, and permitted the use of them, and also of their

Anemone trifolia.

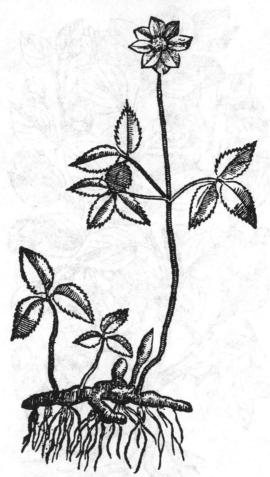

Fig. 41. "Anemone trifolia" (Dodoens,
Pemptades, 1583]

figures, in one another's books. To attempt to ascertain exactly what degree of merit should be attributed to each of the three would be a task equally difficult and thankless.

Jules-Charles de l'Escluse [more commonly known as

Plate viii

CHARLES DE L'ÉCLUSE (1526–1609)

[Print in the Botany School, Cambridge. This engraving is from a copy made by M. Ambrose Tardieu of Paris from a portrait in the possession of M. Rota]

Charles de l'Écluse

Charles de l'Écluse, or Clusius] (pl. viii, f.p. 84) was born in 1526 at Arras, which at that date belonged to Flanders, and, like Dodoens, he passed the closing years of his life at Leyden. As a youth, after studying at various universities, he went to Montpellier. Here he was received into the house of the botanist and physician Guillaume Rondelet, who also at different times numbered F. Platter, M. de l'Obel, P. Pena, J. Bauhin, J. d'Aléchamps, and J. Desmoulins among his pupils. As a botanist, Rondelet is remembered, not by any surviving work, but by the tradition of his gifts as a teacher. He was indolent about writing, and his natural instrument of expression was the spoken word. In this respect he has been compared with Socrates, to whom his portrait bears some physical resemblance. De l'Écluse owed much to Rondelet, who not only cured him of a severe illness, but also confirmed him in that love of botany which was to be, for the rest of his long and nomadic life, his ruling passion. After the Montpellier period, we hear of him in various parts of Europe, at one time writing or translating for Rondelet, Dodoens, or Plantin, and at another time acting as tutor to the sons of some personage of importance. The ill-health which always dogged him, was but one of his troubles. His family, who were of the reformed faith, were persecuted in the name of religion, and one of his near relatives was burnt at the stake. Charles himself was reduced to extreme poverty, through giving up all he had to his father, whose possessions were confiscated. The misfortunes which Charles de l'Écluse suffered, would, indeed, have crushed most men, but he rose above them, finding his happiness in his inexhaustible capacity for work, and in his power of forming friendships, wherever he went, with men of learning. In youth he had sat at the feet of Melanchthon, and in old age J. J. Scaliger was his intimate. He paid more than one visit to England, and came into relation with Sir Philip Sidney, and also with Sir Francis Drake, through whom he obtained plants from the New World.

85

IV. The Botanical Renaissance

De l'Écluse had a reputation for versatility scarcely inferior to that of his contemporary, the "Admirable" Crichton. In addition to his botanical knowledge, he is credited with an intimate acquaintance with Greek, Latin, Italian, Spanish, Portuguese, French, Flemish, German, law, philosophy, history, cartography, zoology, mineralogy, numismatics, and epigraphy. His gift for languages was of great service in the diffusion of botany, for, besides translating the Flemish herbal of Rembert Dodoens into French, he made Latin versions of the Portuguese work of Garcia de Orta, and the Spanish writings of Christoval Acosta and Nicolas Monardes, which we shall consider later.

The first botanical work which de l'Écluse claimed as original, arose out of an adventurous expedition to Spain and Portugal with two pupils, from which he brought back two hundred new species. The description of his finds was published by Plantin in 1576, under the title of *Rariorum aliquot stirpium per Hispanias observatarum Historia*. Woodblocks were engraved purposely for this book (see figs. 42, p. 87; 66, p. 156; 117, p. 231), but, for the confusion of the bibliographer, some of them were used also to illustrate Dodoens' work, in the interval during which the Spanish flora awaited publication. The relation between the botanists of the Low Countries was indeed fully fraternal. To Rembert Dodoens, who had given him a home in the period of his sorest troubles, de l'Écluse felt himself, as he wrote, "united by friendship of old", and, in alluding to Dodoens' prior use of some of his illustrations, he added, "whatever friends possess ought to be freely shared". De l'Écluse's generosity was not mere lip service, but was carried into the practical conduct of life; on the death of his father, he waived the rights of seniority, and thus the hereditary title of Seigneur de Watènes, to which he would have succeeded, passed to his younger brother.

In 1573 de l'Écluse was invited to Vienna by the Emperor

Maximilian II, and he remained there for about fourteen years; for a time he was employed in connection with the

Fig. 42. "Lacryma Iob", *Coix Lachryma-Jobi* L., Job's-tears
[de l'Écluse, *Rariorum...per Hispanias*, 1576]

imperial gardens. From Vienna he was able to explore the mountains of Austria and Hungary, and in 1583 he described the botany of these regions in a book in which he also included an account of various oriental plants which had reached Vienna through Constantinople. Late in life he was appointed to a professorship at Leyden, and he then brought together his Spanish and Hungarian floras, and other works, into a single folio, published in 1601, as *Rariorum plantarum historia*. In the preface to this—his *magnum opus*—he speaks in moving terms of the pleasure that he has taken in observing the delectable variety of plants in different regions of the earth. Botanical discoveries have given him, he says, as much joy as if he had found a prodigious treasure; but at seventy-six, with failing health, he is content to hand on the torch to the younger generation, hoping that they may be excited to similar or greater efforts.

When we come to analyse the contribution which de l'Écluse made to science, we find that his original work was descriptive rather than classificatory; he is said to have added over six hundred to the number of known plants. It is characteristic of his wide mental scope, that his botanical range was not confined, like that of most of the early workers, to flowering plants. Appended to his *Rariorum plantarum historia* is a *Fungorum historia*—the first published monograph of its kind. The thoroughness of de l'Écluse's study of the group is revealed by a codex preserved in the Leyden Library, which contains more than eighty sheets of water-colour drawings of fungi. A fully annotated volume, including a facsimile both of the printed *Fungorum historia* and of the codex, has been issued in recent times; it makes good the claim that de l'Écluse should be honoured as the founder of mycology. The paintings in the codex are admirable, and surprisingly modern in style; they were executed under the direction of de l'Écluse by an artist employed by his great friend and patron, Baron Boldiszár [Balthasar] de Batthyány. This Hungarian nobleman was so enthusiastic a botanist, that he once set a Turkish prisoner at liberty, on the condition that he should provide a ransom in the form of beautiful flowers from his native land.

De l'Écluse and de l'Obel

De l'Écluse did much for horticulture, and special gratitude is due to him for the introduction of the potato into Germany, Austria, France, and the Low Countries. Through his connections with the Mediterranean region and the near east, he brought a number of species into cultivation, including various kinds of *Ranunculus*, *Anemone*, *Iris* and *Narcissus*, as well as other bulbous and tuberous plants. He was the chief founder of the bulb culture which has played so conspicuous a part in the history of the Netherlands, and he fully deserves the title given to him by his friend, Marie de Brimen, Princesse de Chimay, "Le père de tous les beaux Jardins de ce pays."

The planning and arrangement of the botanical garden at Leyden was directed by de l'Écluse. Recently the memory of this association has been perpetuated by the conversion of an adjacent plot of land into an exact reproduction of his garden, as it was in 1594. His assemblage of plants was botanical rather than strictly medical in its scope. This is consistent with the fact that de l'Écluse, though he was a licentiate in medicine, was not a practising physician, and was less preoccupied than most herbalists with the medical aspect of botany. He studied plants for their own sakes, and he was not obsessed with the attempt to identify them with the simples mentioned by the writers of antiquity.

De L'Écluse is a man who well repays close acquaintance. The more one studies his career and work, the more significant does he appear, both as a botanist, and as a man in whom was centred much of the botanical life of his period.

The last of the trio of botanists whom we are now considering is Mathias de l'Obel [de Lobel or Lobelius] who was born at Lille in 1538, and died in England, at Highgate, in 1616. His portrait at the age of seventy-six is shown in pl. ix, f.p. 90. The frontispiece of one of his books bears a device of white-poplars as "armes parlantes", his name having taken its origin, traditionally, from the abele-tree. The *Lobelia*, which

89

was dedicated to him, thus has a name with an odd history, since it has transmigrated from a plant to a man, and then to another plant.

De l'Obel studied at Montpellier under Rondelet, whose high opinion of his pupil was shown by the bequest to him of his botanical manuscripts. In 1559 de l'Obel came to England, where he passed a peaceful period under the prosperous sway of Queen Elizabeth, to whom he dedicated a book in terms of hyperbolic praise. Presently he returned to the Low Countries, where he became physician to William the Silent, a post which he held until the assassination of the Stadtholder. A few years later, de l'Obel returned to England, where he remained for the rest of his life. He seems to have been well received in this country, for he was invited to superintend the medicinal garden at Hackney belonging to Lord Zouche, and he eventually obtained the title of Botanist to James I. He deserves to be remembered for his services to British botany, since he is responsible for more than eighty "first records" of our native plants, among which are the grass-of-Parnassus (*Parnassia palustris* L.), the deadly-nightshade (*Atropa Belladonna* L.), the frogbit (*Hydrocharis Morsus-ranae* L.), and the arrowhead (*Sagittaria sagittifolia* L.).

De l'Obel's chief botanical work was the *Stirpium adversaria nova*, published in 1570,1, with Pierre Pena as joint author. Despite the efforts of various ardent researchers, we know surprisingly little about Pena; it is conjectured that, at an early stage, he forsook botany in favour of medicine. For simplicity we will follow current usage in speaking of their joint work, and of the important theoretical ideas which it incorporated, under de l'Obel's name alone; the possibility is not, however, excluded that Pena played an equal, or even a preponderating part in their combined thought. It is to be hoped that future discoveries may throw light upon the relation of these two botanists.

Unlike de l'Écluse, de l'Obel wrote in Latin which has been

Plate ix

MATHIAS DE L'OBEL (1538–1616)

[Engraving by François Dellarame, 1615. Department of
Prints and Drawings, British Museum]

described as "dur et presque barbare", but his matter made ample amends for the uncouthness of his style. The word "adversaria", in the title of his *magnum opus*, carries the modest meaning of "memoranda". The book is still full of interest for botanists in the south of France, owing to the exactness of its statements about the flora of the Montpellier region. Moreover, it is in this work that the system of classification which has given de l'Obel his reputation, is set forth. The main feature of the scheme is that he distinguishes different groups by the characters of their leaves. He is thus led to make a rough separation between the classes which we now call dicotyledons and monocotyledons. The details of his system will be considered in a later chapter (pp. 176, 177).

In 1576 the work was enlarged, and republished as the *Plantarum seu stirpium historia*; it was also translated into Flemish and appeared in 1581 under the title of *Kruydtboeck*. Immediately after the publication of the Flemish herbal, Plantin brought out its engravings (of which figs. 64, p. 151; 76, p. 172; 116, p. 230, are examples) as an album without text, called *Plantarum...icones*. Although they had also been used to illustrate the herbals of Dodoens and de l'Écluse, these pictures, which numbered over 2000, were now grouped according to de l'Obel's arrangement, which was recognised as the best. Linnaeus makes frequent reference to this album, and its woodcuts are familiar in England, as they were used again in Johnson's editions of Gerard's herbal (1633 and 1636).

The frontispiece of the present book is after a picture painted in 1665 by an artist of the Netherlands—Adrian van Ostade. We may venture to guess that the herbal lying open, beside the physician working in his study, was written by one of the "fraternal triumvirate", who were the founders of botany in Ostade's native country.

In 1670—the last year of the period with which we are here concerned—Nylandt's *Nederlandste Herbarius* appeared at Amsterdam. It is interesting as a survival of the old-fashioned type of herbal, including many familiar sixteenth-century woodcuts.

3. THE HERBAL IN ITALY

The Italian botanists of the renaissance did pioneer work, the importance of which can scarcely be over-estimated. Since the revival of classical culture took its origin in Italy, it is not surprising that the botanists of that country should have been the foremost in the intelligent study of Greek biological literature. Even before the end of the fifteenth century, the works both of Dioscorides and of Theophrastus had been printed by Italian presses, not only in Latin, but also in the original Greek. The botanists of Italy were well placed for identifying the plants described by classical authors, since their own flora was related to that of Greece and the other Mediterranean regions which Theophrastus and Dioscorides themselves had known; so the Italians were not compelled, like the botanists whose homes lay north of the Alps, to distort the facts, before they could accommodate the plants of their native country to the procrustean bed of classical description.

One of the chief commentators and herbalists of the later renaissance was Pierandrea Mattioli [Matthiolus] (fig. 43, p. 93), who was born at Siena in 1501. He was the son of a doctor, and his early life was passed in Venice, where his father was in practice. Pierandrea was destined for the law, but his inherited tastes led him away from jurisprudence to medicine. He practised in several different towns, and became physician, successively, to the Archduke Ferdinand, and to the Emperor Maximilian II. In 1577 he died of the plague. We can realise something of the frightful extent of this scourge, from the large part it plays in the history of even so small a company as the botanists of the renaissance. We have already mentioned (p. 64) that Leonhart Fuchs' medical fame was carried as far as England by his successful treatment of one such epidemic. Gaspard Bauhin witnessed three of these disastrous outbreaks, while he was practising as a physician in Basle; and his brother, Jean, had the terrible experience of

Fig. 43. Pierandrea Mattioli, 1501–77 [Engraving by Philippe Galle, *Virorum Doctorum Effigies*, Antwerp, 1572]

being an official plague doctor in Lyons during an epidemic which killed 5000 persons. Besides Mattioli, two other well-known herbalists, Gesner and Zaluziansky, fell victims to the plague in different countries.

Mattioli's botanical *chef-d'œuvre* was his *Commentarii in sex libros Pedacii Dioscoridis*, the gradual production and improvement of which occupied his leisure hours throughout his life; it was first published in 1544. It was translated into many

Pyra.

Fig. 44. "Pyra", *Pyrus communis* L., Pear
[Mattioli, *Commentarii*, 1560]

languages and appeared in a long series of editions. The success of the work was phenomenal; it is said that 32,000 copies of the earlier editions were sold. The title does not do the book justice, for it contains an account of all the plants known to Mattioli. The early versions had small illustrations

(figs. 44, p. 94; 45, p. 95; 109, p. 223; 110, p. 224), but, later
on, editions with superb, large-scale figures were published

Auena.

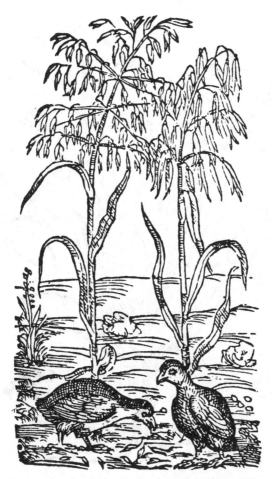

Fig. 45. "Avena", Oats [Mattioli, *Commentarii*, 1560]

in Prague and Venice (figs. 46, p. 96; 47, p. 98; 111, p. 225;
112, p. 226). The estimation in which the *Commentarii* were
held is shown by a clause in the will (dated 1637) of the
diplomatist, Sir Henry Wotton, best remembered now by his

lovely lines on Elizabeth of Bohemia. This clause runs: "I leave to our most Gracious and Vertuous Queen Mary

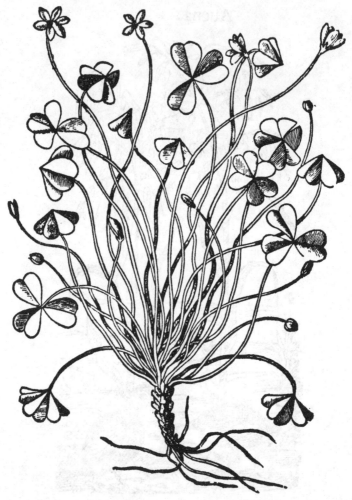

Fig. 46. "*Trifolium acetosum*", *Oxalis* [Mattioli, *Commentarii*, 1565] *Reduced*

[Henrietta Maria, wife of Charles I], *Dioscorides*, with the Plants naturally colored, and the Text translated by *Matthiolo*, in the best Language of *Tuscany*, whence her said

Majesty is lineally descended, for a poor token of my thankful devotion, for the honour she was once pleased to do my private study with her presence."

Mattioli found and noted a certain number of new plants, especially from the Tyrol, but most of the species which he described for the first time, were not his own discoveries, but were communicated to him by others. He placed on record, for instance, the observations made in Turkey by the diplomatist, Busbecq, and his physician, Quakelbeen. In one of his letters Busbecq mentions that he has some botanical drawings that he is keeping for Mattioli, and that he had previously sent him a good many specimens. Moreover, among the manuscripts which Busbecq secured in Constantinople, were two versions of Dioscorides,[1] which he handed over to Mattioli, who refers to this loan in the preface to his *Commentarii* of 1565. Another of Mattioli's correspondents who gave him generous help was Luca Ghini. Like Mattioli, he had projected a botanical work of wide scope, but he allowed his material to be incorporated in the *Commentarii*. The close friendship between these two herbalists is revealed in a letter to Aldrovandi, in which Mattioli writes that the death of Ghini has carried away half his heart.

A later Italian botanist, who also drew his inspiration from Dioscorides, was Fabio Colonna, or, as he is more generally called, Fabius Columna (pl. x, f.p. 98), who was born at Naples in 1567. His father, Girolamo, a well-known man of letters, possessed a library of 2500 volumes, and collected ancient pictures, coins, and statues. Fabio's profession was that of law, but he was also well acquainted with languages, music, mathematics, and optics. He tells us himself that his interest in plants was aroused by his difficulty in obtaining a remedy for epilepsy, a disease from which he suffered. Having tried all sorts of prescriptions without result, he went to the

---

[1] Neither of these was the Anicia Juliana Codex (pp. 8, 9), which seems to have been unknown to Mattioli.

fountain head, Dioscorides, and, after much research, identified valerian as being the herb which that writer had recommended for the purpose. He used it with success; but

Fig. 47. "Malus", *Pyrus Malus* L., Apple [Mattioli, *Commentarii*, 1565] *Reduced*

perhaps, as a modern writer has suggested, we shall be nearer the truth if we attribute the improvement in his health, not to any drug, but to the open air and exercise which

98

Plate x

FABIUS COLUMNA (1567–1650)

[*Ekphrasis*, 1606; probably a self-portrait]

Fig. 48. "Apocynum" [Columna, *Phytobasanos*, 1592]

IV. The Botanical Renaissance

his plant hunting involved, and to the interest of it, which distracted his mind from his bodily ills.

Columna's experience with valerian having convinced him that the knowledge of the plants enumerated by the ancients was in a most unsatisfactory condition, he set himself to produce a work which should make these matters clearer. In his book, which was published in 1592, under the name of *Phytobasanos* (plant touchstone), and in a second volume, *Ekphrasis*, a high standard was reached, both in the descriptions and in the pictures (e.g. figs. 48, p. 99; 124, p. 242).

A passing reference may be made here to certain books, which are botanical accounts of special regions, rather than herbals in the strict sense. One of these is the *Viaggio di Monte Baldo* by Francisco Calzolari, whose name is remembered in the calceolaria beloved of Victorian gardeners. His little work, which was first published in 1566, was among the earliest of the innumerable books which may be classed as local floras, whose history, though it is of real importance, cannot be pursued in these pages.

Another treatise of a more ambitious kind, though also dealing with a special field, is *De plantis Aegypti* by Prospero Alpino, which was published at Venice in 1592 (see fig. 49, p. 101). The author was a doctor, who accompanied the Venetian consul, Giorgio Emo, to Egypt, and who took full advantage of this botanical opportunity. He is said to be the first European writer to mention the coffee plant, which he saw growing at Cairo. Prospero Alpino eventually succeeded to the oldest chair of botany in Europe—that founded at Padua by the Venetian republic in 1533; he enriched the garden of his university with Egyptian plants. This garden was created in 1542, and is thus the doyen among existing botanic gardens. It is a happy illustration of the continuity of botany, that the very palm-tree, which inspired Goethe's theory of metamorphosis, and of which he treasured the dried leaves as fetishes in after years, was already growing in the Padua

Fig. 49. "Kalli", *Salicornia*, Glasswort [Prospero Alpino,
De plantis Aegypti, 1592]

garden during Alpino's professorship, and that it still flourishes there to-day. A convincing portrait of Alpino, which has been ascribed, though without proof, to his friend Leandro Bassano, is preserved in the Botanical Institute at Padua; it is reproduced here on a small scale (pl. xi).

An Italian savant who deserves mention, though he published nothing of importance, is Aldrovandi (1522–1605), who in 1550 founded at Bologna an extensive museum of natural history, probably the most ancient of its kind in Europe; specimens of fishes, which he had acquired under the influence of Rondelet, whom he met at Rome, formed the original nucleus of the museum. Mathias de l'Obel, and Jean and Gaspard Bauhin, attended the botanical lectures of Aldrovandi. He was a plant collector for fifty years, and was liberal in his communications to his friends, so that his influence on the progress of the science was considerable. The volumes belonging to his library, which are still preserved at Bologna, bear the revealing inscription, "Ulyxis Aldrovandi et Amicorum".

There is yet another botanist of sixteenth-century Italy who must be included in our brief historical survey, not because his work was of real value, but because it was widely popular. This was Castor Durante, a physician who issued botanical compilations bedizened with Latin verse. His principal book is the *Herbario Nuovo*, published at Rome in 1585 (figs. 50, p. 103; 77, p. 174; 121, p. 236). A certain German volume, *Hortulus sanitatis*, seems to be a translation from Durante, but its origin is not wholly clear. As an illustration of this author's pleasingly unscientific manner, we may cite his account of the "Arbor tristis". The picture which accompanies it (fig. 50, p. 103), shows, beneath the moon and stars, a tree, the trunk of which has a human shape. The description, as it occurs in the *Hortulus sanitatis*, may be rendered as follows:

"Of this tree the Indians say, there was once a very

Plate xi

PROSPERVS ALPINVS
PROF. SIMPLICIVM HORTIQ. PRAEFECTVS
AB ANNO MDCIII. AD MDCXVI.

PROSPERO ALPINO (1553–1617)

[From a painting which has been attributed to Leandro Bassano, in the collection belonging to the Botanic Garden at Padua]

beautiful maiden, daughter of a mighty lord called Parisa-taccho. This maiden loved the Sun, but the Sun forsook her because he loved another. So, being scorned by the Sun, she slew herself, and when her body had been burned, according to the custom of that land, this tree sprang from her ashes. And this is the reason why the flowers of this tree shrink so intensely from the Sun, and never open in his presence. And

Fig. 50. "Arbor Malenconico" or "Arbor Tristis", Tree-of-sorrow [Durante, *Herbario Nuovo*, 1585]

thus it is a special delight to see this tree in the night time, adorned on all sides with its lovely flowers, since they give forth a delicious perfume, the like of which is not to be met with in any other plant, but, no sooner does one touch the plant with one's hand, than its sweet scent vanishes away. And however fair the tree has appeared, and however sweetly it has bloomed at night, directly the Sun rises in the morning, it not only fades, but all its branches look as though they were withered and dead."

IV. The Botanical Renaissance

4. THE HERBAL IN SPAIN AND PORTUGAL

In the sixteenth century, Spanish versions of Dioscorides, and the commentaries of the Portuguese Jew, Amatus Lusitanus, first appeared in print. The Spaniards and Portuguese, however, made their special contribution to botany rather as travellers, who recorded the plants of the distant lands to which their spirit of adventure carried them. In the fifteen hundreds, the kingdom of Portugal formed the connecting link between Europe and India. Vasco da Gama had reached Calicut on the Malabar coast by the sea route in 1498, and from this time onwards, for a hundred years or more, commerce with the east was almost entirely in the hands of his countrymen. Goa Dourada, "Golden Goa", fell to Albuquerque in 1510, and became the capital of Portuguese India. To this city, in 1534, the physician, Garcia de Orta [ab Horto, ab Orto, or del Huerte] set sail from the Tagus, reaching India after a voyage lasting six months. He was well equipped for useful work in India, since he had studied medicine at more than one Spanish university, and had been a lecturer at Lisbon. At Goa he practised as a physician with marked success, and amassed a fortune. The poet Camoens was one of his friends and addressed a sonnet to him. After experience in the use of eastern medicaments extending over a quarter of a century, he wrote a work called *Coloquios dos simples, e drogas he cousas mediçinais da India.* It was published at Goa in 1563, and was thus one of the first European books to be printed in India. It holds a special place in botanical history, because it is cast in the form of a dialogue in which Orta, to a certain extent, typifies the "Arabist", who accepts the teachings of classical writers, not simply as they stand, but as corrected and amplified by Avicenna and the mediaeval Arab physicians; his interlocutor, on the other hand, tends towards the opinions of the "Hellenists", who discard the experience of all intermediate generations, and regard the

Greeks as the sole authorities. De Orta, having established himself in a country so far from Spain, was unusually free to express his own opinions; he writes, indeed, in one of the *Colloquies*—"even I, when in Spain, did not dare to say anything against Galen or against the Greeks".

The *Colloquies* give an early glimpse of a number of oriental products, such as cloves, mace and nutmeg, ginger, cinnamon, assafoetida, and betel-nut. Although the aim of the book is utilitarian, we meet with an occasional note on something of purely botanical import. For instance there is a brief description of a plant with sensitive leaves—which appears to be an oxalid, *Biophytum sensitivum* Dec.—and also a mention of the sleep position of the leaves of the tamarind. The *Colloquies* are, indeed, well worth studying, and it is fortunate that a modern version now exists in English.

De Orta's own work was unillustrated, but in 1578 a volume with pictures, the text of which was little more than a Spanish translation of the *Colloquies*, was published at Burgos under the title of *Tractado de las drogas y medicinas de las Indias Orientales con sus Plantas*. One of the illustrations is shown in fig. 51, p. 106. Christoval Acosta, a native of Burgos, was responsible for this book; his passion for travel had led him as far as India, where he had become acquainted with de Orta. Both the *Colloquies* and Acosta's tractate reached a wider public through Latin versions made by de l'Écluse, who had fresh figures drawn to illustrate de Orta's work. The copy of the *Colloquies* which belonged to de l'Écluse is still extant; it is annotated in Latin, in a script which, though almost microscopic, is perfectly clear and elegant.

Just as we owe our knowledge of the plants of the East Indies primarily to the Portuguese, so it is from the Spaniards that we have our first acquaintance with the botany of the New World. Nicolas Monardes, who was born at Seville in 1493, placed on record some of the earliest discoveries in a little book published in two parts in 1569 and 1571, and in a more complete form in 1574. The title-page of the first instalment is reproduced in fig. 52, p. 107. In 1577 the work was translated into English by John Frampton. This English

Fig. 51. "Piper Nigrum", Pepper [d'Aléchamps, *Historia generalis plantarum*, vol. II, 1587, after "Pimienta negra", Acosta, *Tractado de las drogas*, 1578]

version, from which our quotations are taken, either has a title-page which calls it *The Three Bookes* of Monardes, or it appears under the more picturesque name of *Joyfull newes out of the newe founde worlde.* Monardes points out that the existence of differences between plants of different regions

ℒ DOS LIBROS, EL V-

NO QVE TRATA DE TODAS LAS COSAS
que traen de nueſtras Indias Occidentales, que ſiruen
al vſo de la Medicina, y el otro que trata de la
Piedra Bezaar, y de la Yerua Eſcuerçonera.
Cõpueſtos por el doctor Nicoloſo de Monardes Medico de Seuilla.

IMPRESSOS EN SEVILLA EN CASA DE
Hernando Diaz, en la calle de la Sierpe.
Con Licencia y Priuilegio de ſu Mageſtad.
Año de 1 5 6 9.

Fig. 52. Title-page [Monardes, *Dos libros*, 1569]

107

was known to the Aristotelian school, and adds, "And as there is discovered newe regions, newe kyngdomes, and newe Provinces, by our Spanyardes, thei have brought unto us newe Medicines and newe Remedies." We realise the close connection that then existed between England and Spain, from the translator's remark that "the afore saied Medicines ...are now by Marchauntes and others brought out of the West Indias into Spaine, and from Spain hether into Englande, by suche as dooeth daiely trafficke thether". John Frampton himself had been in commerce in Spain, and his dedication, to "Maister Edward Dier Esquire" (the author of "My mynde to me a kyngdome is"), begins with the words: "Retournyng right worshipfull, home into Englande out of Spaine, and now not pressed with the former toiles of my old trade. I to passe the tyme to some benefite of my countrie, and to avoyde idlenesse: tooke in hande to translate...the thre bookes of Doctour Monardes...." Frampton's version is pleasant for its fresh descriptions of things that are now familiar to us. The sunflower, or "Hearbe of the Sunne" is, he writes, "a straunge flower, for it casteth out the greatest flowers, and the moste perticulars that ever hath been seen, for it is greater then a great Platter or Dishe, the whiche hath divers coulers...it showeth marveilous faire in Gardines"

Monardes gives one of the earliest pictures of tobacco to appear in a printed book (fig. 53, p. 109); Frampton substituted a better figure, published at about the same time by Pena and de l'Obel (fig. 116, p. 230). Monardes may not always have discriminated between tobacco and coca, for he tells us that the Negroes and Indians after inhaling tobacco smoke "doe remaine lightened, without any wearinesse, for to laboure again: and thei dooe this with so great pleasure, that although thei bee not wearie, yet thei are very desirous for to dooe it: and the thyng is come to so muche effecte, that their maisters doeth chasten theim for it, and doe burne the *Tabaco*, because thei should not use it".

In the present century, *Joyfull Newes* has been republished as one of a series of Tudor translations.

The Spaniards were remarkable for the interest and respect which they accorded to the pharmacopoeia of the "Indians" of

EL TABACO. 3

Fig. 53. "El Tabaco", *Nicotiana Tabacum* L., Tobacco
[Monardes, *Segunda parte del libro,* 1571]

Mexico. From a book by Hernandez, physician to Philip II, which appeared in 1615, it is clear that the doctors of Spain adopted a great many herbs, under Mexican names, into their own *materia medica*. This interest in the native culture had, indeed, come into play even before Monardes had published his pioneer work. In the Vatican Library there is an illustrated manuscript herbal, composed in 1552, at the Roman Catholic College of Santa Cruz, Mexico, by two Aztecs—Martin de la Cruz, who is described as an "Indian physician...who is not theoretically learned, but is taught only by experience", and Juannes Badianus, who translated the work into Latin. The *Badianus herbal*, as it is called, is now available to students in a facsimile version issued in the United States.

5. THE HERBAL IN SWITZERLAND

Among the many men of learning whose names are associated with Switzerland, one of the most renowned is Konrad Gesner [Conrad Gessner] (pl. xii), who was born at Zurich in 1516, the son of a poor furrier.

His taste for botany was due, in the first instance, to the influence of his mother's uncle, a Protestant preacher, with whom he lived as a boy, and who taught him the names of the flowers in his garden, which was small, but filled with plants of all kinds. In Paris, where Konrad was sent to study, the richness of the libraries, and the delight of associating with learned men, led him into a course of omnivorous, and perhaps rather superficial, reading, for which he blamed himself in later life; but catholicity of taste was part of his natural equipment, and he probably did best by ranging widely. After an interval of school teaching at Zurich, he betook himself to Basle, where he entered more methodically upon the study of medicine, at the same time attempting to support himself by working at a Greek and Latin lexicon. His student life was interrupted by a period as professor of Greek at Lausanne, after which he again turned to medicine. While he was at Lausanne, he compiled a catalogue of plant names in Greek, Latin, German, and French.

Gesner owed much to his native town of Zurich, which more than once allotted to him a "Reisestipendium", thus

Plate xii

KONRAD GESNER (1516–1565)
[Print in the Botany School, Cambridge]

enabling him to study in France and elsewhere. After he took his doctorate, his relation with his birthplace became even closer, for he was appointed first to the professorship of Philosophy there, and then to that of Natural History, which he held until he died of the plague in his fiftieth year.

Gesner was more remarkable for encyclopaedic versatility than for critical scholarship; he has been called the Pliny of his age. He wrote on bibliographical and linguistic subjects, as well as on medicine, mineralogy, zoology, and botany. The botanical works published during his life were not of great importance, but, at the time of his death he had already prepared a large part of the material for a general history of plants, which was intended as a companion work to his *Historia animalium.* In order to illustrate it, he had collected 1500 drawings; the majority were original, while the rest were founded on previous woodcuts, especially those of Fuchs. The undertaking was so far advanced that some of the figures had been drawn upon the wood, and certain blocks had even been engraved. Shortly before his death, Gesner handed over the whole collection, and the manuscripts, to his friend Kaspar Wolf, to be prepared for publication. Wolf seems to have made an honest effort to carry out Gesner's wishes, but he was hampered by weak health, and the task, as a whole, proved to be beyond his powers. He succeeded only in issuing a few of the woodcuts as an appendix to Simler's *Vita Conradi Gesneri* (e.g. fig. 54, p. 112). Finally he sold everything to Joachim Camerarius the Younger, who brought a number of Gesner's figures before the public, but only by the indirect method of using them, associated with blocks of his own, to illustrate works of which he was author or editor.

About two hundred years after the death of Gesner, his drawings and blocks came into the possession of the bibliographer, C. J. Trew. In the middle of the eighteenth century, a proportion of them, from Trew's library, were published by Schmiedel, thus giving Gesner his due, so far as it was possible at that late date. Trew's collection of Gesner's material (including many figures besides those which

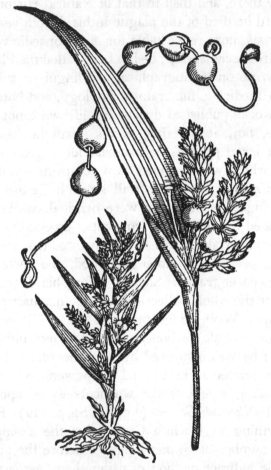

Fig. 54. "Lachryma Iob", *Coix Lachryma-Jobi* L.,
Job's-tears [Simler, *Vita Conradi Gesneri, 1566*]

Schmiedel issued, as well as numerous letters) was eventually acquired by the University of Erlangen. For more than a century the collection was lost to sight, but recently part of it has been recovered in the Erlangen Library and studied with pious care. It is to be hoped that Gesner's remains, as a whole, may eventually be retrieved and published.

It is clear from Gesner's letters that his interest in botany was thoroughly scientific. If he had placed more of his work on record in print, he would probably shine as a discoverer of new species, for his figures indicate that he was acquainted with a number of plants which are now credited to de l'Écluse, Gaspard Bauhin, and others, as first describers.

Among Gesner's numerous scientific correspondents was Jean Bauhin, a brilliant young man, twenty-five years his junior. Their acquaintance began when Bauhin, who was born in 1541, was only about nineteen. Gesner's opinion of him is evidenced by his mode of address. He writes to him as "Erudito et singularis spei Iuveni Ioanni Bauhini", or "Ornatissimo et doctissimo iuveni".

Jean Bauhin was the son of a French physician, Jean Bauhin the Elder, a native of Amiens, who had been converted to protestantism by reading Erasmus' Latin translation of the New Testament. His change of faith exposed him to religious persecution, which he avoided by retreating to Switzerland, where his sons, Jean and Gaspard, were born. Both of them entered their father's profession. The medical tradition was, indeed, remarkably strong among the Bauhins; for six generations, extending over two hundred years, an unbroken succession of members of the family were medical men.

After Jean Bauhin had studied for a time at the University of Basle, he went to Tübingen, where he learned botany from Leonhart Fuchs, and ate at his table. From Tübingen he proceeded to Zurich, and accompanied Gesner on some excursions in the Alps. After further travel on his own account, and a period at the University of Montpellier, he reached Lyons, where he came to know d'Aléchamps, and undertook to assist him with his *Historia generalis plantarum.* Bauhin began to occupy himself with this work, but difficulties arose on account of his religion, and he was obliged to quit France.

Jean Bauhin's chief botanical work, the *Histoire universelle des plantes*, was an ambitious undertaking, which he did not live to see published. However, in 1619, seven years after his death, a preliminary sketch of the *Histoire* was brought out by his son-in-law, Cherler; and, in 1650,1, the *magnum opus* itself appeared. It was edited by Chabrey (Chabraeus) and called *Historia plantarum universalis*. Planned on a large scale, it included descriptions of 5000 plants.

Jean Bauhin's more famous brother, Gaspard [Caspar] (pl. xiii), was born in 1560, and was thus the younger by nearly two decades. Gaspard was a delicate and backward child, who did not learn to speak plainly until his fifth year, but he seems to have outgrown these handicaps. As a student he worked at Basle, Padua, Bologna, Montpellier, Paris and Tübingen. He had intended to visit other German towns, but, on account of his father's ill-health, he was obliged to return to Basle. Here he took his doctor's degree, which he received from the hands of Felix Platter. After some years he was appointed to a professorship in anatomy and botany, and in 1614, on Platter's death, he succeeded to the professorship of medicine.

Inspired by the example of his brother, he conceived the plan of collecting into a single work everything that was known about systematic botany. His extensive early travels served as a good preparation for this task, since he had not only observed and collected widely, but had established relations with the best botanists all over Europe. He formed a herbarium of about 4000 plants, including species from remote countries. This collection is still preserved in the University of Basle. Besides study bearing directly upon his great project, he did a considerable amount of critical and editorial work, which also had its value in relation to his main plan. He produced new editions of Mattioli's *Commentarii*,

Plate xiii

GASPARD BAUHIN (1560–1624)

[Engraving by J. T. de Bry, *Theatrum Anatomicum*, 1605]

and of the herbal of Tabernaemontanus, and published a criticism of d'Aléchamps' *Historia generalis plantarum.*

There is a marked parallelism between the botanical

Fig. 55. "Solanum tuberosum esculentum", Potato
[Bauhin, *Prodromos*, 1620]

careers of the Bauhin brothers. The main part of Gaspard's chief work never saw the light at all, but his son brought out one instalment of it, many years after his father's death. Gaspard was, however, more fortunate than Jean, in that he lived to see the publication of three important preliminary

115   8-2

volumes giving the result of his researches: the *Phytopinax* (1596), the *Prodromos theatri botanici* (1620), and the *Pinax theatri botanici* (1623). The figure of the potato from the *Prodromos* is reproduced in fig. 55, p. 115; the plant still retains Bauhin's name in the binomial form—*Solanum tuberosum*.

The *Pinax* of 1623 is Gaspard Bauhin's main work. By this date, owing to the number of different names bestowed upon the same species by different authors, and the varying identifications of the herbs described by the ancients, plant nomenclature and synonymy had reached a condition of extreme confusion. Bauhin's *Pinax* deserved its title (πίναξ, chart, or register), since it contained a complete and methodical concordance of the names of plants. It brought order out of chaos, and its author has been held in honour as "législateur en botanique". The *Pinax*, which dealt with about 6000 species, was recognised as pre-eminent for many years. Morison, Ray, and Pitton de Tournefort, to a great extent retained its nomenclature. Linnaeus, in 1730, while still at Uppsala University, received a copy of the second edition of the *Pinax* as payment for giving lessons in botany to a fellow medical student. He evidently used his copy constantly from that time, and it contains about 3000 determinations added by him in the margins. Gaspard Bauhin's work was thus carried on into modern botany.

6. THE HERBAL IN FRANCE

The most important French botanist of the earlier renaissance period was Jean Ruel, or Joannes Ruellius, as he is more commonly called (1474–1537). He was a physician, and a professor in the University of Paris, and devoted himself chiefly to the study of classical botany. He did good service by his Latin translation of Dioscorides (1516), which was used by Mattioli in his *Commentarii*. He also wrote a general botanical treatise on Aristotelian lines, *De Natura stirpium* (1536).

Addideris vocem , fuerit Dalechampius ipſe
Expreſſa ad viuum, cuius imago fuit.

Fig. 56. Jacques d'Aléchamps, 1513–88 [Enlarged from a woodcut,
1600 circa, Department of Prints and Drawings, British Museum]

IV. The Botanical Renaissance

In the later renaissance of the sixteenth and early seventeenth centuries, France does not seem at first sight, to have made a very large contribution towards the development of

Fig. 57. "Ornithogalum magnum" [d'Aléchamps, *Historia generalis plantarum,* 1586,7]

the herbal. It must be remembered, however, that the school of botany at Montpellier, under Rondelet, was second to none in Europe; and also that Jean and Gaspard Bauhin, and the publisher, Christophe Plantin, were French by extraction, though Switzerland and the Netherlands were their countries by adoption. Most of the principal herbals published in other languages appeared in French versions quite early in their history, sometimes in a modified form; Antoine du Pinet, for example, translated the *Commentarii* of Mattioli, while *L'Histoire des plantes*, by Geofroy Linocier, was founded, in part, upon the works of Mattioli and Fuchs. A herbalist whose name stands higher than that of du Pinet or of Linocier is Jaques d'Aléchamps [Daléchamps] (fig. 56, p. 117), who was born at Caen in 1513. After studying medicine at Montpellier, he entered upon the practice of it at Lyons, where he remained until his death in 1588. In 1586,7, the *Historia generalis plantarum*, a book which formed a compendium of much of the botany of the late sixteenth century, was published at Lyons; it commonly goes by the name of *Historia plantarum Lugdunensis*. D'Aléchamps is not mentioned upon the title-page, but there is evidence for assigning the senior authorship to him, although Jean Bauhin and Jean Desmoulins [Molinaeus] also had a considerable share in the work.

7. THE HERBAL IN ENGLAND

The pioneer name among English herbalists of the renaissance period is that of William Turner[1], physician and divine, the "Father of British Botany". He was a north-countryman, a native of Morpeth in Northumberland, where he was born probably between 1510 and 1515. He received his education at what is now Pembroke College, Cambridge. Pembroke deserves to be held in special honour by biologists, for a hundred years later, Nehemiah Grew, one of the greatest British botanists of the seventeenth century, also became a student at this college.

Like many herbalists of that time, William Turner was closely associated with the Reformation. He embraced the views of his friends and instructors at Cambridge, Nicholas

---

[1] This account of Turner's life is taken from Jackson, B.D. (1877); Appendix II.

119

IV. The Botanical Renaissance

Ridley and Hugh Latimer, and fought for the reformed faith throughout his life, both with pen and by word of mouth. A ban was put upon his writings in the reign of Henry VIII, and for a time he suffered imprisonment, but, when Edward VI came to the throne, his fortunes improved, and, after a long and tedious period of waiting for preferment, he obtained the Deanery of Wells. Difficulty in ejecting the previous Dean caused much delay in obtaining possession of the house, and Turner lamented bitterly that, in the small and crowded temporary lodging, "i can not go to my booke for ye crying of childer & noyse y^t is made in my chamber".

During Mary's reign Turner was a fugitive, and the former Dean of Wells was reinstated. However, when Elizabeth ascended the throne, the position was reversed, and Turner came back to Wells, "the usurper", as he calls his rival, being ejected. But his triumph was short-lived, for in 1564 he was suspended for nonconformity. His controversial methods were violent in the extreme, and he seems to have been a thorn in the flesh of his superiors. The Bishop of Bath and Wells wrote on one occasion that he was "much encombred wth m^r Doctor *Turner* Deane of *Welles*, for his undiscrete behavior in the pulpitt: where he medleth wth all matters, and vnsemelie speaketh of all estates, more than ys standinge withe discressyon".

Christian doctrine was by no means the only subject that engaged Turner's attention. He had taken a medical degree either at Ferrara or Bologna, and, in the reign of Edward VI, he was physician to the Duke of Somerset, the Protector. He had travelled much in Italy, Switzerland, Holland and Germany, at the periods when his religious opinions excluded him from England. One of the great advantages which he reaped from his wanderings was the opportunity of studying botany at Bologna under Luca Ghini, whom he describes as one of his "maisters in Italye". Another savant with whom he became acquainted on the Continent, was

Konrad Gesner, whom he visited at Zurich, and with whom he maintained a warm friendship. He corresponded also with Leonhart Fuchs.

Turner's earliest botanical work was the *Libellus de re herbaria novus* (1538). This was followed, ten years later, by another little work, *The names of herbes in Greke, Latin, Englishe Duche and Frenche wyth the commune names that Herbaries and Apotecaries use.* Both these rarities are now available in facsimile. In the preface to *The names of herbes*, Turner tells us that he had projected a Latin herbal, and had indeed written it, but refrained from publishing it because, when he "axed the advise of Phisicianes in thys matter, their advise was that I shoulde cease from settynge out of this boke in latin tyll I had sene those places of Englande, wherein is moste plentie of herbes, that I might in my herbal declare to the greate ho[n]oure of our countre what numbre of sovereine and strang herbes were in Englande that were not in other nations, whose counsell I have folowed deferryng to set out my herbal in latin, tyl that I have sene the west countrey, which I never sawe yet in al my lyfe, which countrey of all places of England, as I heare say is moste richely replenished wyth all kindes of straunge and wonderfull workes and giftes of nature, as are stones, herbes, fishes and metalles". He explains that while waiting to complete his herbal, he has been advised to publish this little book in which he has set forth the names of plants. He adds, "and because men should not thynke that I write of that I never sawe, and that Poticaries shoulde be excuselesse when as the ryghte herbes are required of them, I have shewed in what places of Englande, Germany and Italy the herbes growe and maye be had for laboure and money".

Turner's *chef-d'œuvre* was his *Herball*, published in three instalments, the first in London in 1551, the first and second together at Cologne in 1562, during his exile in the reign of Mary, and the third part, together with the preceding, in

IV. The Botanical Renaissance

1568. The title of the first part runs as follows, *A new Herball, wherin are conteyned the names of Herbes . . . with the properties degrees and naturall places of the same, gathered and made by Wylliam Turner, Physicion unto the Duke of Somersettes Grace.* The figures illustrating the herbal are, for the most part, those in the octavo edition of Fuchs' work, published in 1545.

The dedication of the herbal, in its completed form, to Queen Elizabeth, throws some light on Turner's life, and incidentally on that illustrious lady herself. The doctor recalls, with pardonable pride and perhaps a touch of blarney, an occasion on which the Princess Elizabeth, as she then was, had conversed with him in Latin. "As for your knowledge in the Latin tonge," he writes, "xviii. yeares ago or more, I had in the Duke of Somersettes house (beynge his Physition at that tyme) a good tryal thereof, when as it pleased your grace to speake Latin unto me: for although I have both in England, lowe and highe Germanye, and other places of my longe traveil and pelgrimage, never spake with any noble or gentle woman, that spake so wel and so much congrue fyne and pure Latin, as your grace did unto me so longe ago."

Turner defends himself against the insinuation that "a booke of wedes or grasse (as some in despite of learninge will call precious herbes) is a righte unmete gift for such a Prince"; and certainly, if we accept his account of the state of knowledge at the time, the need for a herb book in English must have been most urgent. He explains that, while he was still at Cambridge, he endeavoured to learn the names of plants, but, "suche was the ignorance in simples at that tyme", that he could get no information on the subject, even from physicians. He claims that his herbal has considerable originality. In his own words—"they that have red the first part of my Herbal, and have compared my writinges of plantes with those thinges that Matthiolus, Fuchsius, Tragus, and Dodoneus wrote in yᵉ firste editiones of their Herballes, maye easily perceyve that I taught the truthe of certeyne plantes, which these above named writers either knew not at al, or ellis erred in them greatlye.... So y^t as I learned something of them, so they ether might or did learne somthinge of me agayne, as their second editions maye testifye. And because I would not be lyke unto a cryer y^t cryeth a loste horse in the marketh, and telleth all the markes and tokens that he hath, and yet never sawe the horse, nether coulde knowe the horse if he sawe him: I wente into Italye and into diverse partes of Germany, to knowe and se the herbes my selfe".

His herbal contains many evidences of Turner's independence of character, which appears to have been greater than his independence of thought. He fought against what he regarded as superstition in science with the same ardour with which he entered upon religious polemic. The legend of the human form of the mandrake receives scant mercy at his hands. As he points out, "The rootes which are conterfited and made like litle puppettes and mammettes, which come to be sold in England in boxes, with heir, and such forme as a man hath, are nothyng elles but folishe feined trifles, and not naturall. For they are so trymmed of crafty theves to mocke the poore people with all, and to rob them both of theyr wit and theyr money. I have in my tyme at diverse tymes taken up the rootes of Mandrag out of the grounde, but I never saw any such thyng upon or in them, as are in and upon the pedlers rootes that are comenly to be solde in boxes." Turner was, however, by no means the first to dispute the mandrake superstition; in *The grete herball* of 1526 it is definitely refuted—"for nature never gave forme or shape of mankynde to an herbe"; moreover it is ignored in some works that are of even earlier date. The hoax was long-lived, for we find Gerard exposing it again at the end of the sixteenth century.

Turner delighted in pouring scorn upon any superstitious notions which he detected in the writings of his contemporaries, and seems to have been particularly pleased if he could show that in any disputed matter they were wrong, while the ancients, for whom he had a great reverence, were right. For instance he has much to say about a theory, held by Mattioli, in opposition to the opinions of Theophrastus and Dioscorides, that the broomrape (*Orobanche*) could kill other plants merely by its baneful presence, without any physical contact. He declares that this view is contrary to reason, authority and experience, and points out that the figure which Mattioli gives is faulty, in omitting to show the roots, which are the real instruments of destruction. He triumphantly concludes, "And as touchynge experience, I know that the freshe and yong Orobanche hath commyng out of the great roote, many lytle strynges...wherewith it taketh holde of the rootes of the herbes that grow next unto it. Wherefore Matthiolus ought not so lyghtly to have defaced the autorite of Theophrast so ancient and substantiall autor." Despite Turner's respect for the classical botanists, he realised that there were gaps in their knowledge; the third part of his herbal is, indeed, occupied with plants "whereof is no mention made nether of yᵉ old Grecianes nor Latines".

Turner's herbal is arranged alphabetically, and does not show evidence of any interest in the relationships of the plants. It is as individuals, and essentially as "simples", that he regarded them. It is usual to treat British systematic botany as dating from his work, which includes the first scientific records of no less than 238 of our native plants. Turner's botanical achievement seems the more remarkable, when we realise that plants were not the only group that he studied. He wrote about birds, and also contributed information concerning British fishes to Gesner's *Historia animalium*.

In 1578, twenty-seven years after the appearance of the first part of Turner's herbal, an English translation, from de l'Écluse's French version, of Dodoens' *Crüydeboeck* of 1554, was printed at Antwerp, but published in London. This translation was made by Henry Lyte, who, about 1529, had been born of an ancient stock, being the thirteenth lineal descendant of that name and family—a family which still flourishes, and can count among its modern representatives the Rev. H. F. Lyte, author of "Abide with me", and Sir Henry Maxwell-Lyte, the historian.

The title of Lyte's book is as follows: *A Nievve Herball, or Historie of Plantes: wherein is contayned the whole discourse and perfect description of all sortes of Herbes and Plantes: their divers and sundry kindes: their straunge Figures, Fashions, and Shapes: their Names, Natures, Operations, and Vertues: and that not onely of those which are here growyng in this our Countrie of Englande, but of all others also of forrayne Realmes, commonly used in Physicke. First set foorth in the Doutche or Almaigne tongue, by that learned D. Rembert Dodoens, Physition to the Emperour: And nowe first translated out of French into English, by Henry Lyte Esquyer.* Lyte's own copy of de l'Écluse's version, with numerous notes in a small and delicate hand, varied with red and black ink, and endorsed on the title-page—"Henry Lyte taught me to speak Englishe"—is preserved in the British Museum. This copy proves that Lyte was no mere mechanical translator, for the herbal is annotated and corrected, references to de l'Obel and Turner being inserted. At the beginning of the book, there is a long set of doggerel verses, "in commendation of this worke", from which one gathers that, after the English translation of the *Histoire des plantes* was finished, Rembert Dodoens himself sent extra material, which was incorporated in *A Nievve Herball.* The relevant stanza runs as follows:

> Great was his toyle, whiche first this worke dyd frame.
> And so was his, whiche ventred to translate it,
> For when he had full finisht all the same,
> He minded not to adde, nor to abate it.
> But what he founde, he ment whole to relate it.
> Till *Rembert* he, did sende additions store.
> For to augment *Lytes* travell past before.

IV. The Botanical Renaissance

The illustrations used in *A Nievve Herball* were, in the main, those which had appeared in the French translation by de l'Écluse, and were thus, for the most part, ultimately

Rha. Reubarbe.

Fig. 58. "Reubarbe", *Centaurea Rhaponticum* L. [Lyte, *A Nievve Herball*, 1578, after Dodoens, *Crüÿdeboeck*, 1554]

derived from Fuchs, though some blocks (e.g. "Reubarbe", fig. 58) had been added to Fuchs' set by Dodoens.

In the year following the publication of Lyte's book, Edmund Spenser's *Shepheardes Calender* appeared; the *Aprill*

126

Spenser and A Nievve Herball

eclogue, with its lyrical praise of Queen Elizabeth, includes a
stanza enumerating a series of flowers:

> Bring hether the Pincke and purple Cullambine,
> With Gelliflowres:
> Bring Coronations, and Sops in wine,
> worn of Paramoures.
> Strowe me the ground with Daffadowndillies,
> And Cowslips, and Kingcups, and loued Lillies:
> The pretie Pawnce
> And the Cheuisaunce
> Shall match with the fayre flowre Delice.

In general, Spenser's allusions to flowers are of a conventional and literary type, and seem to indicate that he himself
felt no particular interest in plants; there is, indeed, nothing
quite comparable with this stanza elsewhere in his poetry.
Now, when we turn to Lyte's herbal, we find that the five
flower names that Spenser first mentions occur within
sixteen pages, as well as "pances", and the wallflower,
which may, in the opinion of the present writer, be identified
with Spenser's "Cheuisaunce". It is not, perhaps, too wild a
speculation that Spenser availed himself of a charming set of
plant names that met his eye in glancing through *A Nievve
Herball*, and that he was influenced by the page here reproduced (fig. 59, p. 128), in which "Carnations", "Gillofers", and "Soppes in wine" are illustrated. There is one
circumstance which may be held to increase the probability
that Spenser handled *A Nievve Herball*. Sir Philip Sidney's
sister, Lady Mary Herbert, Countess of Pembroke, to whom
Spenser addressed poems, lived at no great distance from
Henry Lyte, who offered his herbal to Queen Elizabeth,
"From my poore house at Lytes-carie within your Maiesties
Countie of Somerset." That Lady Mary cared for flowers is
suggested by the fact that Jacques Le Moyne dedicated *La
Clef des Champs* to her. It thus seems not unlikely that she
would have acquired a copy of the herbal, and that it might
have been shown to Spenser.

Of Gillofers.   Chap.bij.

✤ The Kyndes.

Ader the name of Gillofers (at this time) diuerse sortes of floures are contayned. wherof they call the first the Cloue gillofer whiche in deede is of diuerse sortes & variable colours: the other is the small or single Gillofer & his kinde. The third is that, which we cal in English sweete williams, & Colminiers: whereunto we may well iopne the wilde Gillofer or Cockow floure, which is not much unlike the smaller sort of garden Gillofers.

Vetonica altilis.

Carnations, and the double-cloaue Gillofers.

Vetonica altilis minor.

The single Gillofers, Soppes in wine, and Pinkes, &c.

✤ The Defcription.

'The Cloue gillofer hath long small blades, almost like Leeke blades. The stalke is round, and of a foote and halfe long, full of ioyntes and knops, & it beareth

Fig. 59. "Gillofers", Carnations and Pinks [Lyte, *A Nievve Herball*, 1578] *Reduced*

128

John Gerard

After Henry Lyte, we next come to John Gerard (pl. xiv, f.p. 130), the best known of all the English herbalists, but who does not, it must be confessed, fully deserve the fame which has fallen to his share. Gerard (whose name is more correctly spelt without the final "e", which it assumes on the title-page of his herbal) was a Barber-Surgeon, but his energies appear to have been employed chiefly upon horticulture. For twenty years he had a renowned garden in Holborn—then a fashionable district—and also supervised those of Lord Burleigh in the Strand, and at Theobalds in Hertford-shire. In 1596 he issued a list of the plants which he cultivated in Holborn; this list has the distinction of being the first professedly complete catalogue ever published of the contents of a single garden.

Gerard's reputation rests however on a much larger work, *The Herball or Generall Historie of Plantes*, published by John Norton in 1597; but the manner in which this book originated does the author little credit. It seems that Norton had commissioned a certain Dr Priest to translate Dodoens' final work, the *Pemptades* of 1583, into English, but Priest died before the work was finished. Gerard adopted Priest's translation, completed it, altering the arrangement from that of Dodoens to that of de l'Obel, and published it as his own. There is a remark in his address to the reader at the beginning of his herbal, which can only have been a deliberate lie: "Doctor *Priest*, one of our London Colledge, hath (as I heard) translated the last edition of *Dodonaeus*, which meant to publish the same; but being prevented by death, his translation likewise perished." After the manner of the period, the herbal is embellished with a number of prefatory letters, in one of which, written by Stephen Bredwell, a statement occurs which is so inconsistent with Gerard's own remarks that he certainly committed an oversight in allowing it to stand. In Bredwell's words—"D. *Priest* for his transla-tion of so much as *Dodonaeus*, hath hereby left a tombe for his honorable sepulture. Master *Gerard* comming last, but not the least, hath many waies accommodated the whole worke unto our English nation."

The *Herball* is a massive volume in clear Roman type, which gives the book a modern appearance when it is compared with the black letter pages of Turner or Lyte. It contains about 1800 woodcuts, of which only a very small number appear to be new. One of these, which represents the potato (fig. 67, p. 158), is perhaps the first figure of the plant ever published. The early history of this vegetable is obscure, and Gerard's belief that it was a native of Virginia is only one of the many errors entangled with the subject. Nearly all the illustrations in the *Herball* are from the blocks which had been used by Tabernaemontanus in his *Eicones* of 1590. Unfortunately Gerard's botanical knowledge was not sound enough to enable him to couple these wood-blocks with their appropriate descriptions, and de l'Obel was requested by the printer to correct the author's blunders. This he did, according to his own account, in very many places, but yet not so many as he wished, since Gerard became impatient, and stopped the process of emendation, partly on the ground that de l'Obel had forgotten his English. After this episode, the relations between the two botanists seem, not unnaturally, to have become somewhat strained.

The value of Gerard's work must inevitably be at a discount, when we realise that it is impossible, from internal evidence, to accept him as a credible witness. His oft-quoted account of the "Goose tree", "Barnakle tree", or the "tree bearing Geese", removes what little respect one may have felt for him as a scientist—not, indeed, because he held an absurd belief, which was widely current at the time, but because he described it, with utter disregard of truth, as confirmed by his own observations. To illustrate the origin of the geese he gives a figure (fig. 60, p. 131), which is not, however, new. Gerard relates that trees, actually bearing

130

Plate xiv

JOHN GERARD (1545–1607)

[*The Herball*, 1636]

shells, which open and hatch out barnacle geese, are said to occur in the northern parts of Scotland and the "Orchades",

The breede of Barnakles.

Fig. 60. "The breede of Barnakles" [Gerard, *The Herball*, 1597]

but he frankly states that on this point he does not possess first-hand knowledge. He proceeds, however, to remark, "But what our eies have seene, and hands have touched, we

131   9-2

shall declare. There is a small Ilande in Lancashire called the Pile of Foulders, wherein are found the broken peeces of old and brused ships, some whereof have beene cast thither by shipwracke, and also the trunks or bodies with the branches of old and rotten trees, cast up there likewise: whereon is found a certaine spume or froth, that in time breedeth unto certaine shels, in shape like those of the muskle, but sharper pointed, and of a whitish colour; wherein is conteined a thing in forme like a lace of silke finely woven, as it were togither, of a whitish colour; one ende whereof is fastned unto the inside of the shell, even as the fish of Oisters and Muskles are; the other ende is made fast unto the belly of a rude masse or lumpe, which in time commeth to the shape and forme of a Bird: when it is perfectly formed, the shel gapeth open, and the first thing that appeereth is the foresaid lace or string; next come the legs of the Birde hanging out; and as it groweth greater, it openeth the shell by degrees, till at length it is all come foorth, and hangeth onely by the bill; in short space after it commeth to full maturitie, and falleth into the sea, where it gathereth feathers, and groweth to a foule, bigger then a Mallard, and lesser than a Goose."

It is interesting to compare Gerard's attitude towards the barnacle goose story with that of William Turner, who wrote about fifty years earlier. Turner makes no pretence of vouching for the tale from his own experience, but he accepts it because he has it on what he regards as good authority—and respect for authority which he believed to be good was one of Turner's most marked characteristics. His account has been translated in the following words:

"When after a certain time the firwood masts or planks or yard-arms of a ship have rotted on the sea, then fungi, as it were, break out upon them first, in which in course of time one may discern evident forms of birds, which afterwards are clothed with feathers, and at last become alive and fly. Now lest this should seem fabulous to anyone, besides the common evidence of all the long-shore men of England, Ireland, and Scotland, that renowned historian Gyraldus,...bears witness that the generation of the Bernicles is none other than this. But inasmuch as it seemed hardly safe to trust the vulgar and by reason of the rarity of the thing I did not quite credit Gyraldus,...I took counsel of a certain man, whose upright conduct, often proved by me, had justified my trust, a theologian by profession and an Irishman by birth, Octavian by name, whether he thought Gyraldus worthy of belief in this affair. Who, taking oath upon the very Gospel which he taught, answered that what Gyraldus had reported of the generation of this bird was absolutely true, and that with his own eyes he had beholden young, as yet but rudely formed, and also handled them, and, if I were to stay in London for a month of two, that he would take care that some growing chicks should be brought in to me."[1]

There were many variants of the goose tree legend: in one of them, recorded by Hector Boethius [Boece] in his sixteenth-century Scottish chronicle, the geese, as in Gerard's tale, grow from driftwood, "in the small boris and hollis" of which "growis small wormis. First thay schaw their heid and feit, and last of all they schaw thair plumis and wyngis. Finaly quhen thay ar cumyn to the iust mesure and quantite of geis, thay fle in the aire, as othir fowlis dois."[2]

The credulity of the herbalists who accepted the goose tree legend is the more surprising when we realise that, as early as the thirteenth century, Albertus Magnus had disproved it by observing that barnacle geese were hatched from eggs like other birds; on this, as on so many other subjects, Albertus

---

[1] *Turner on Birds:...first published by Dr William Turner, 1544.* Edited by A. H. Evans, Cambridge, 1903, p. 27. [The original passage will be found in *Avium praecipuarum...Per Dn. Guilielmum Turnerum,...Coloniae excudebat Ioan. Gymnicus, 1544.*]

[2] Hector Boethius, *Heir beginnis the hystory and croniklis of Scotland...Translatit laitly in our vulgar and commoun langage, be maister Johne Bellenden...And Imprentit in Edinburgh, be me Thomas Davidson* [1536] (Cap. xiv of the "Cosmographie").

was in advance of his age. It is refreshing to find that Fabius Columna, in his *Phytobasanos* (1592), flatly denies the truth of the legend, and that it is rejected in the later versions of Gerard's herbal, published after his death. It reappears, however, at as recent a date as 1783 in the last version of the old Egenolph herbals, known as *Adam Lonicers Kräuter= Buch*; and it may be that a shadow of the myth still lingers in our speech, when we call an unfounded rumour a *canard*.

The first edition of Gerard's herbal held the field without a competitor for more than a generation. It was not until it began to be noised abroad that a certain John Parkinson would soon produce a new herbal to take its place, that the successors of Gerard's original publisher were brought to the point of undertaking a second edition. In 1632 they commissioned Thomas Johnson, a well-known London apothecary and botanist, to carry out the work, with the proviso that it must be completed within the year. This heavy task Johnson accomplished with marked success, even adding a balanced and comprehensive historical introduction. He recalls Gaspard Bauhin in his scholarly anticipation of modern methods of editorship; he has, for example, a system of marking the text to distinguish the degrees to which he has altered or rewritten Gerard's descriptions. Johnson's new version was illustrated with a set of 2766 blocks, previously used in the botanical books published by Plantin. The *Herball*, thus transformed, reached a far higher level than Gerard's own edition.

Though it is as the editor of Gerard that Johnson is most frequently remembered, his independent botanical work is, in reality, of special significance. His accounts of his plant hunting excursions, published from 1629 onwards (p. 284), are the earliest attempts to record all the plants of England and Wales, with their localities. But for his early death, there is little doubt that he would have carried out his intention of writing, in conjunction with his friend Goodyer, a descriptive and illustrated flora, of a type which did not, in fact, come into being until the eighteenth century.

We know so little of Johnson's personal life that any intimate touch is welcome, so it is pleasant to come upon a letter from Sir Henry Wotton, addressed "To my verie loving and learned frend Mr. Johnson, Apothecarie, at his howse on Snowe Hill, London", asking where could be bought "one of your Gerrards, well and strongly bound: Next, where I may have for my monye, all kinde of coloured Pynkes to sett in a Quarter of my Garden or any such flowers as perfume the Ayre".

In the Civil War, Johnson fought on the Royalist side, and died from the effects of a shot wound, while taking part in the defence of Basing House. He was described as "no lesse eminent in the Garrison for his valour and conduct, as a Souldier, then famous through the Kingdom for his excellency as an Herbarist, and Physician".

John Parkinson (1567–1650) was the last British writer of the period which we are considering who belonged to the true lineage of herbalists—though he was, in some ways, a degenerate representative. His portrait is shown in pl. xv, f.p. 136. Like Gerard he cultivated a famous garden in what is now the heart of London. Besides Gerard's in Holborn, and Parkinson's in Long Acre, other well-known London gardens of the period were John Tradescant's at Lambeth, and Mistress Tuggy's at Westminster, celebrated for the "excellencie and varietie" of her "Gillofloures, Pinkes, and the like".

Parkinson was honoured with the title of Herbarist to Charles I. The earlier of the two books, by which he is remembered, was rather of the nature of a gardening work than of a herbal. It appeared in 1629 under the title, *Paradisi in Sole Paradisus Terrestris. A Garden of all sorts of pleasant flowers which our English ayre will permitt to be noursed up;... together With the right orderinge planting and preserving of them and their uses and vertues.* The words "Paradisi in Sole" form a pun upon the author's name. The book—which was dedicated to Queen Henrietta Maria, with the prayer that she would accept "this speaking Garden"—is now accessible in a facsimile reprint.

The preface to this work is entirely at variance with the idea that scientific knowledge has been acquired only gradually by the human race. In Parkinson's words: "God, the Creator of Heaven and Earth, at the beginning when he created *Adam*, inspired him with the knowledge of all naturall things (which successively descended to *Noah* afterwardes, and to his Posterity): for, as he was able to give names to all the living Creatures, according to their severall natures; so no doubt but hee had also the knowledge, both what Herbes and Fruits were fit, eyther for Meate or Medicine, for Use or for Delight."

Elaborate directions for the setting and treatment of a garden precede an account of a large number of plants cultivated at that time, with some mention of their uses. The book is illustrated with full-page wood-engravings of no great merit, in each of which a number of different plants are represented; fig. 61, p. 137, is taken from part of one illustration. Some of the figures are original, while others are copied from the books of de l'Écluse, de l'Obel and others.

As a sequel to this work, Parkinson issued a much larger volume, dealing with plants in general, and called the *Theatrum botanicum: the theater of plants, or, an herball of a large extent.* According to the preface of the *Paradisus Terrestris*, the author's original intention was merely to supplement his description of the Flower Garden by an account of "A Garden of Simples". This scheme grew into one of a broader nature, but without losing its predominantly medical character. Now and again, despite his comparatively late date, Parkinson's imagination is touched with mediaevalism. He is eloquent on the subject of that rare and precious

Plate xv

JOHN PARKINSON (1567–1650)

[*Theatrum botanicum*, 1640]

commodity, the horn of the unicorn, which is a cure for many bodily ills. He describes the animal as living, "farre remote from these parts, and in huge vast Wildernesses among other most fierce and wilde beasts". Nevertheless, despite occasional outbreaks of this type, Parkinson's herbal shows certain improvements upon that of Gerard and Johnson. He took advantage of Bauhin's *Pinax*, and was thus able to give the

Fig. 61. "*Oxyacantha seu Berberis*. The Barbary bush" [Part of a large woodcut, Parkinson, *Paradisus Terrestris*, 1629]

nomenclature of each plant in detail. Many of de l'Obel's manuscript notes are also inserted; but the scheme of classification adopted is markedly inferior to that of de l'Obel.

Some of the most interesting of our native plants were first noticed by Parkinson. He is responsible, for instance, for the earliest records of the Welsh-poppy (*Meconopsis cambrica* Vig.), the strawberry-tree (*Arbutus Unedo* L.), and the ladies'-slipper (*Cypripedium Calceolus* L.).

England was now well supplied with herbals and we may

IV. The Botanical Renaissance

contrast Turner's lament about Cambridge in the sixteenth century (p. 122) with an entry in Abraham de la Pryme's diary, in 1690, when he was a student at St John's College, in that university: "In this my fresh-man's year, by my own propper studdy and industry, I got the knowledge of all herbs, trees, and simples, without any body's instruction or help, except that of herbals: so I could know any herb at first sight."

After the time of Parkinson we pass into a very different epoch, in which the greatest names are those of Robert Morison (b. 1620) and John Ray (b. 1627 or 1628). Their chief works appeared after the close of the period selected for special study in this book, and, as they were botanists in the modern sense, rather than herbalists, we will not attempt any discussion of their writings.

While Morison and Ray were advancing the subject of systematic botany, the Italian, Marcello Malpighi (b. 1628), and the Englishman, Nehemiah Grew (b. 1641), were laying the foundations of the science of plant anatomy. Their work, also, is outside our present scope, and it is only mentioned at this point in order to show that the latter part of the seventeenth century witnessed a considerable revolution in the science. From this period onwards, with the initiation of new lines of enquiry, the importance of the herbal steadily declined, and, although books coming under this heading have been produced even in the twentieth century, the era of their pre-eminence has long been overpast.

8. THE ORIGIN OF HERBARIA

In the early days of the botanical renaissance, progress was slower than it need have been, because none of the herbalists adopted the simple expedient of pressing and drying specimens, as a means of recording and communicating discoveries, and as a basis for comparison between plants which flourish in different localities and at different seasons. To-day,

Plate xvi

PIERRE QUTHE (b. 1519) and his herbarium
[From a portrait, at the age of 43, by François Clouet in the
Musée du Louvre. Cl. Archives photographiques—Paris]

when we are familiar with herbaria, it is difficult to realise that, until the sixteenth century had well begun, the making of such collections was not part of the routine of botanical work. We know that, from a much earlier date, individual plants had occasionally been preserved by pressing; moreover the existence of a thirteenth-century recipe for retaining the colour of dried flowers suggests that continued research may carry the origin of herbaria further back into the past. So far as we can tell from the present evidence, however, the Italian botanist, Luca Ghini (?1490–1556), seems to have been the sole initiator, in the renaissance period, of the art of herbarium making, which was then disseminated over Europe by his pupils. It has been suggested that, during the mediaeval and earlier ages, the expense of such materials as papyrus, parchment, and paper, would have made the pressing of plants impracticable; but it has been pointed out, on the other hand, that this argument is not altogether convincing, since thin sheets of wood, or pieces of textiles, would have served the purpose tolerably well. It seems that we must be content to attribute the belated origin of herbaria, and their diffusion from a single centre, to the humiliating fact that amongst mankind the inventive spirit is rare, while the spirit of imitation is universal.

Luca Ghini was professor of botany at Bologna, and also worked for a time at Pisa. He published nothing, and it is as a teacher and a correspondent that he is known to us. It is on record that he sent dried plants gummed upon paper to Mattioli in 1551, and that, at about that period, he possessed some three hundred pressed specimens. It seems, however, that his herbarium, which is now lost to us, must have been in existence long before 1551, since the oldest extant example is that of his pupil, Gherardo Cibo, who apparently began to collect at least as early as 1532. It is to be hoped that future discoveries will pierce the obscurity that still surrounds these early phases in the history of collections of dried plants.

IV. The Botanical Renaissance

The first printed reference to a herbarium appears to be due to Amatus Lusitanus, who, in his exposition of Dioscorides (1553), mentions that an Englishman, John Falconer, a man after his own heart ("vir mea sententia"), possessed a collection of dried plants. Since Amatus remarks that the plants were sewn and gummed "with wonderful art" ("miro artificio") into a book, we may conclude that he had not previously seen a herbarium. The same collection is mentioned in 1562 by William Turner, as "maister Falkonners boke". That Turner's reference is to a herbarium, and not to a printed book, is proved by the fact that he names Falconer elsewhere as one of those learned Englishmen, who have published nothing, although they "have as muche knowledge in herbes, yea, more than diverse Italianes and Germanes, whyche have set furth in prynte Herballes and bokes of simples". Falconer had travelled in Italy, and it is probable that he, also, learned the art of herbarium-making from Ghini, either directly or indirectly. Turner, Aldrovandi, and Cesalpino—who were, all three, pupils of Ghini—also formed herbaria in the middle years of the sixteenth century. Aldrovandi seems to have been the first to aim at a collection of dried plants, which should include those of the whole world. The value of herbarium specimens as a source for pictures of plants of distant countries, was soon realised; Mattioli mentions the practice of soaking them before drawing, in order to restore their natural appearance.

The collections so far mentioned have been either Italian or English; other early examples are those of Konrad Gesner in Switzerland; of Caspar Ratzenberger, who began a herbarium in 1556, while he was still a student at Wittenberg; and of Jehan Girault, whose collection, dated 1558, is still extant in Paris. In different cities of Europe, more than twenty herbaria are preserved, which were formed, or at least begun, in the sixteenth century.

A recent find, which has been fully examined, is that

of the herbarium of Felix Platter, the eminent physician of Basle, who lived from 1536 to 1614. He studied at Montpellier, where, according to his diary of 1554, he collected "viler kreuter, die ich in papier zierlich inmarkt". It is probable that Rondelet, Platter's teacher, had acquired the art of drying plants from Ghini. The whereabouts of Platter's herbarium had long been unknown, and it was supposed that it must have suffered destruction: a historian of herbaria, writing more than forty years ago, lamented, "Quelles étaient les plantes de l'herbier de Plater, quel en était le nombre? Nous l'ignorons." Now, however, a happy discovery has brought to light eight folio volumes of this herbarium among the treasures of the University of Bern. These volumes, though probably representing less than half of the original collection, yet include as many as 813 species drawn from a wide area—Switzerland, Italy, France, Spain, and Egypt. The dried plants are well preserved and arranged, and some of them have retained their colour admirably. Platter must have regarded a colour record as essential, for, in some species of campanula, the difficulty that the corollas turn brown on drying, has been met by replacing them by imitations cut out of larkspur flowers.

Great as is the intrinsic interest of Platter's herbarium, some will feel that its association value is even greater, since Michel, Seigneur de Montaigne, examined it with pleasure when he passed through Basle in 1580, the very year in which the first edition of the *Essais* was published. Platter's "livre de simples" was evidently a new toy to Montaigne. He recorded in his journal that "au lieu que les autres font pindre les herbes selon leurs coleurs, lui a trouvé l'art de les coler toutes naturelles si proprement sur le papier, que les moindres feuilles et fibres y apparoissent come elles sont". He noted with surprise that the pages could be turned over without the plants dropping out, and that some of these were actually more than twenty years old.

IV. The Botanical Renaissance

Information about the making of herbaria must have been transmitted by word of mouth alone for more than seventy years, for it was apparently not until the publication in 1606 of Adrian Spieghel's *Isagoges*, that detailed instructions for forming a collection of dried plants became available in print. Spieghel, a native of Brussels, finally occupied the chair at Padua. In his *Isagoges*—a general treatise on botany—he explains the method of pressing between sheets of good paper, under gradually increasing weights, and notes that the plants must be examined and turned over daily. When they are dry, they are to be laid upon inferior paper ("charta ignobilior"), and, with brushes of graded sizes, painted with a special gum, for which he gives the recipe. The plants are then to be transferred to sheets of white paper; linen is to be laid over them, and rubbed steadily until they adhere to the paper. Finally the sheets are to be placed between paper, or in a book, and subjected to pressure until the gum dries.

Spieghel realises the great importance of herbaria, and observes that the labour expended in making one is deserving of high praise. He himself calls such a collection a "Winter garden" ("Hortus hyemalis"), while other early writers call it "lebendig Kreuterbuch", "Herbarius vivus", or "Hortus siccus". The word "Herbarium", used in the modern sense, makes its first appearance in print—so far as the present writer is aware—in Pitton de Tournefort's *Elemens* of 1694.

In England herbaria cannot have been common, even in the latter part of the seventeenth century, for until Samuel Pepys saw that of John Evelyn, such collections seem to have been unknown to him—and it was seldom that anything curious escaped his enquiring mind. He writes on 5 November 1665: "By water to Deptford, and there made a visit to Mr Evelyn....He showed me his Hortus Hyemalis; leaves laid up in a book of several plants kept dry, which preserve colour, however, and look very finely, better than any Herball."

In 1763 Adanson, in his methodical way, enumerated four

causes to which he ascribed the progress of botany, and he honoured "Les Erbiers" (which he also calls "Jardins vivans même pendant l'hyver") by including them in this short list of encouraging circumstances; the other three were: "La protection des Souverains et des Grands", "Les Voiages favorisés", and "L'Établissement des Jardins de Botanike".

By good fortune we possess, in François Clouet's sympathetic portrait of his friend, Pierre Quthe, a contemporary representation of a sixteenth-century apothecary, with his herbarium open beside him. This picture is reproduced, on a small scale, in pl. xvi, f.p. 138.

9. THE REVIVAL OF ARISTOTELIAN BOTANY

The subject of Aristotelian botany scarcely comes within the scope of a book on herbals, but, at the same time, it cannot be separated sharply from the botany of the herbalists. We have already shown that, in the middle ages, Albertus Magnus developed the tradition of Aristotle and Theophrastus. In the renaissance period, there was again a revival of this aspect of the study, as well as of herbal botany. The first work of importance in this line was the *De Natura stirpium* of Jean Ruel (1536), but in the present brief summary we cannot discuss it. We must pass on to the year 1583, in which the Italian savant, Andrea Cesalpino (pl. xvii, f.p. 144), produced a great work under the title of *De plantis libri xvi*. This volume is, in part, closely related to the herbals, since much of it is concerned with plant description; Linnaeus evidently studied it with care, for, in his personal copy, he added his own generic names in the margins throughout. The fame of *De plantis* does not, however, rest upon the descriptive sections, but upon the first book, which contains an account of the theory of botany, which is basically Aristotelian. Cesalpino's strength lay in the fact that he approached his subject with a trained mind; he had learned the lesson which Greek thought had then, and has now, to offer to the scientific worker—the lesson of how to think. His work suffered, however, from the defects of his qualities, and his reverence for the classics led him into an over-literal and inelastic acceptance of Aristotelian doctrines. The chief tangible contribution which he made to botanical science, was his insistence upon the prime importance of the reproductive organs. This was the idea on which he chiefly laid stress in his system of classification, to which we shall return in a later chapter.

A botanist who had something in common with Cesalpino was the Bohemian author, Adam Zaluziansky von Zaluzian (1558–1613). The work by which he is known is the *Methodi herbariae, libri tres,* which was published at Prague in 1592. It opens with a survey of botany in general, which is of interest as showing an approach to the modern scientific standpoint, in so far as the author pleads for the treatment of botany as a separate subject, and not as a mere branch of medicine. His remarks on this point may be translated as follows: "It is customary to connect Medicine with Botany, yet scientific treatment demands that we should consider each separately. For the fact is that in every art, theory must be disconnected and separated from practice, and the two must be dealt with singly and individually in their proper order before they are united. And for that reason, in order that Botany, which is, as it were, a special branch of Natural Philosophy [*Physica*], may form a unit by itself before it can be brought into connection with other sciences, it must be divided and unyoked from Medicine."

A later writer, in France, who concerned himself with the theory of botany, was Guy de la Brosse, physician to Louis XIII, who published in 1628 a book, *De la nature, vertu, et utilité des plantes,* dedicated to "Monseigneur le tres-illustre et le tres-reverand Cardinal Monseigneur le Cardinal de Richelieu". De la Brosse would have been shocked to find his work classed under Aristotelian botany, since he himself constantly inveighed against the authority of the classics;

Plate xvii

ANDREA CESALPINO (1519–1603)

[Drawn by G. Zocchi and engraved by F. Allegrini, 1765, after an old portrait in the Museum of the Botanic Garden at Pisa. Print in the Botany School, Cambridge]

but he was deeply influenced by the attitude of the Aristotelians, even when he disagreed with them. This is shown in his attraction towards such problematic topics as variation within individual species, the sensitiveness of plants, and the nature of their souls. Although de la Brosse's work is a strange medley, including much childish speculation, he sometimes faintly foreshadows the theoretical biology of to-day, which—like systematic botany—has drawn its inspiration from the mind of ancient Greece.

Chapter V

THE EVOLUTION OF THE ART OF PLANT DESCRIPTION

robably one of the chief objects, which the early herbalists had in view in writing their books, was to enable the reader to identify simples used in medicine. Nevertheless, until the sixteenth century was well advanced, the pictures in the herbals were often so conventional, and the descriptions so inadequate, that it must have been an almost impossible task to arrive at the names of plants by their aid alone. The idea which suggests itself is that a knowledge of the actual plants was transmitted by word of mouth, and that, in practice, the herbals were used only as reference books, from which to learn the healing qualities of herbs with whose appearance the reader was already familiar. If this supposition is correct, it perhaps accounts for the very primitive state in which the art of plant description remained during the earlier period of the botanical renaissance.

When we turn to the Aristotelian school, we find that the writings of Theophrastus include certain plant descriptions, which, although they seem somewhat rudimentary when judged by modern standards, are remarkable in their way, and far better than those in the first printed herbals. The mediaeval Aristotelian, Albertus Magnus, was also decidedly original in his accounts of flowers, in which attention was drawn to various points which failed to catch the eye of many writers more recent in date. For instance, in describing the flower of the borage he distinguished the green calyx, the corolla with its ligular outgrowths, the five stamens and the central pistil, though naturally he failed to understand their functions. He observed that, in the lily, the calyx was absent, but that the petals themselves showed transitions from green to white. He noticed the early fall of the calyx in the poppy, and its persistence until the ripening of the fruit in the rose. He also pointed out that the successive whorls of sepals and petals alternated with one another, and concluded that this was a device for the better protection of the flower.

Albertus revealed a certain appreciation of flower structure in general, by his classification of floral forms under three types:

(1) Bird-form (e.g. columbine, violet and dead-nettle).

(2) Pyramid- or Bell-form (e.g. convolvulus).

(3) Star-form (e.g. rose).

When we leave the Aristotelian botanists, and turn to those who studied the subject primarily from the medical point of view, we find a definite falling off in the power of description. The accounts of the plants in *De materia medica* of Dioscorides, for example, are, as we have already pointed out, so brief and meagre that exact identification is generally difficult and, not infrequently, impossible.

The herbal of Apuleius Platonicus, which played an important part in the middle ages, scarcely makes any attempt at describing the plants with which it deals. Such a paragraph as the following, from the translation of an Anglo-Saxon manuscript of the eleventh century to which we have already referred, gives an account of a plant, which, compared with the other descriptions in this codex, may fairly be called precise and full:

"This wort, which is named radiolus, by another name everfern, is like fern; and it is produced in stony places, and in old house steads; and it has on each leaf two rows of fair spots, and they shine like gold."

V. Plant Description

The group of late fifteenth-century herbals which we discussed in Chapter II—the *Latin* and the *German Herbarius* and the *Ortus sanitatis*—are alike in giving very short and inadequate accounts of the characters of the plants enumerated, although their descriptions often have a certain naïve

Fig. 62. "Cardamomum", ? *Solanum dulcamara* L., Bittersweet [*Ortus sanitatis*, 1491]

charm. It is scarcely worth while to give actual examples of their methods. It will perhaps suffice to quote a few specimens from the 1529 edition of *The grete herball*, which is a direct descendant of these works, though in date it belongs to the sixteenth century. The wood-sorrel, for instance, is dealt with as follows: "This herbe groweth in thre places, and

148

specyally in hedges, woodes and under wallessydes and hath leves lyke. iii. leved grasse and hath a soure smell as sorell, and hath a yelowe flowre." The expression "yelowe flowre" is an indication of the continental origin of *The grete herball,* for it cannot apply to our British *Oxalis acetosella* L., which has a delicate white bell veined in pink. As another example we may cite chicory, which is described as having "croked and wrythen stalkes, and the floure is of ye colour of the skye". Of the water-lilies, we receive a still more generalised account: "Nenufar is an herbe that groweth in water, and hath large leves and hath a floure in maner of a rose, the rote thereof is called treumyan and is very bygge. It is of two maners. One is whyte, and another yelowe." Occasionally we meet with a hint of more detailed observation. For instance, the coloured central flower in the umbel of the carrot is mentioned, though in terms that sound somewhat strange to the modern botanist. We read that it "hath a large floure and in the myddle therof a lytell reed prycke".

The descriptions in Banckes' *Herball* (1525), though they show a slight but distinct superiority over those in *The grete herball,* still leave much to the imagination. We may take two examples, chosen as being among the best in the book. Of "*Apium ranarum*" (now called *Ranunculus sceleratus* L.) we are told: "This herbe...men call water crowfote. This herbe hathe yelow flowres, as hathe crowfote and of the same shappe, but the leves are more departed, it hath a long stalke, and out of that one stalke groweth many stalkes small by the sydes. This herbe groweth in watry places." Of shepherd's-purse (*Capsella Bursa-pastoris* Medic.) the herbalist writes: "This herbe hathe a small stalke and full of braunches and ragged leves and a whyte flowre. The coddes therof be lyke a purse."

The *Herbarum vivae eicones* of Otto Brunfels (1530) is the first herbal illustrated with drawings which are consistently true to nature. The descriptions, on the other hand, are

V. Plant Description

quite unworthy of the figures, being mostly borrowed from earlier writers. The excellence of the wood-blocks, with which the German Fathers of Botany enriched their books, was, at first, an actual hindrance to the development of the

PIONIA

Fig. 63. "Pionia", Peony [*Latin Herbarius* (Arnaldus de Villa Nova, *Tractatus de virtutibus herbarum*), 1499]

art of plant description. The botanist found that the pencil of the draughtsman could represent every subtlety in the characteristic form of the plant in a way with which no "word painting" could compete. It must, then, have seemed a work of supererogation to set a precise verbal description beneath the drawing.

150

Jerome Bock [Hieronymus Tragus], the next great worker
after Brunfels, was unable to afford illustrations for the

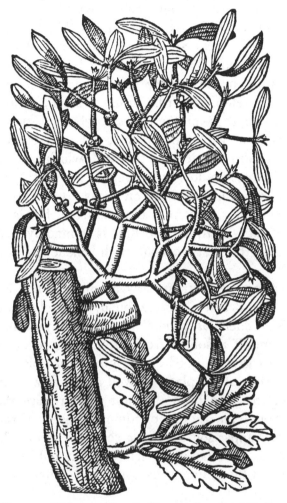

Fig. 64. "Marentacken", *Viscum album* L., Mistletoe
(on Oak) [de l'Obel, *Kruydtboeck*, 1581]

first edition of his herbal. This proved to be a blessing in
disguise, since it stimulated him to deal with plant form as
expressively as possible in words. His descriptions of flowers
and fruits are sometimes excellent, and the way in which he

indicates the general habit is often masterly. As an example we may quote his account of mistletoe plants, which may be translated as follows: "They grow almost in the shape of a cluster, with many forks and articulations. The whole plant is light green, the leaves are fleshy, plump and thick, larger than those of the box. They flower in the beginning of spring, the flowers are however very small and yellow in colour, from them develop, towards autumn, small, round white berries very like those on the wild gooseberry. These berries are full inside of white tough lime, yet each berry has its small black grain, as if it were the seed, which however does not grow when sown, for, as I have said above, the mistletoe only originates and develops on trees. In winter missel thrushes seek their food from the mistletoe, but in summer they are caught with it, for bird-lime is commonly made from its bark. Thus the mistletoes are both beneficial and harmful to birds."

As indicating the contrast between Brunfels and Bock, we may notice that Brunfels, having found, as he tells us, that the ancients are "silent as fishes" about the lily-of-the-valley, considers himself absolved from attempting to describe it, whereas Bock gives a good and original word picture of the vegetative parts of the plant, as well as the flower and fruit. Bock seems to have been the first to introduce the word "pistil", though his use of "pistillum" (pestle) is rather for purposes of comparison than as a scientific term. The word "petal" came into botanical language at a later date; it was suggested by Columna in 1592, and defined as "floris folium".

De historia stirpium, the Latin herbal of Leonhart Fuchs, includes a full glossary of the technical terms used, which is of historic importance, as the first document of the kind in botanical literature. His definitions, however, incline to be nebulous, and are of little scientific value; it will suffice to translate two examples to show their style:

Bock, Fuchs, and Turner

"*Stamens* are the points [apices] that shoot forth in the middle of the flower-cup [calyx]: so called because they spring out like threads from the inmost bosom of the flower."

"*Pappus*, both to the Greeks and to the Latins, is the fluff which falls from flowers or fruits. So also certain woolly hairs which remain on certain plants when they lose their flowers, and afterwards disappear into the air, are pappi, as happens in Senecio, Sonchus and several others."

There is little that is original in Fuchs' descriptions. The account of the male butterbur (fig. 65, p. 154), for instance, in the German edition of his herbal, is taken almost word for word from Bock. It runs: "The flower of butterbur is the first to appear, before the plant or leaves. The flower is cluster-shaped, with many small, pale pinkish flowerets, and is like a fine bunch of vine flowers in full bloom to look at. This large cluster-shaped flower has a hollow stalk, at times a span high; it withers and decays without fruit together with the stalk. Then the round, gray, ash-coloured leaves appear, which are at first like coltsfoot, but afterwards become so large that one leaf will cover a small, round table. They are light green on one side, and whitish or gray on the other. Each leaf has its own brown, hairy and hollow stem, on which it sits like a wide hat or a mushroom turned over. The root grows very thick, is white and porous inside, and has a strong, bitter taste."

Our English herbalist, William Turner, had a special gift for vivid description. He compares the dodder (*Cuscuta*) to "a great red harpe stryng", and the seed vessels of shepherd's-purse to "a boyes satchel or litle bagge". Of the dead-nettle he says: "Lamium hath leaves like unto a Nettel, but lesse indented about, and whyter. The downy thynges that are in it like pryckes, byte not, ye stalk is four-square, the floures are whyte, and have a stronge savor, and are very like unto litle coules, or hoodes that stand over bare heades. The sede is

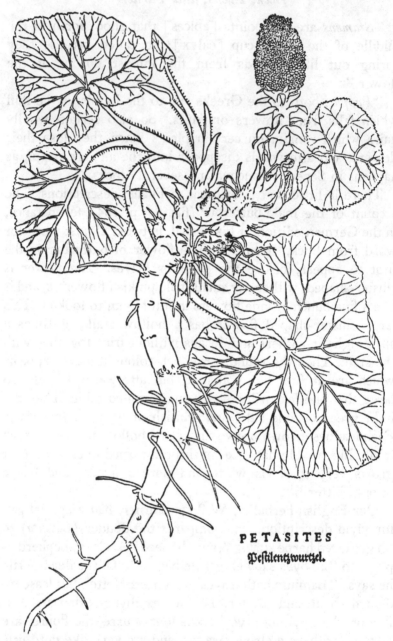

PETASITES

Peſtilentʒwurtʒel.

Fig. 65. "Petasites", *Tussilago Petasites* L., Butterbur (male plant)
[Fuchs, *De historia stirpium*, 1542] *Reduced*

154

blak and groweth about the stalk, certayn places goyng betwene, as we se in horehound."

Henry Lyte's *Nievve Herball* of 1578 was, as we have already mentioned, a translation of the *Histoire des plantes*, which is itself a version by de l'Écluse of the Flemish herbal of Dodoens. We may thus fairly illustrate the style of plant description of the latter herbalist by a quotation from Lyte, since it has the advantage of retaining the sixteenth-century atmosphere, which is too evanescent to cling to a modern translation. Of the snapdragon he writes: "The great Antirrhinon hath straight round stemmes, and full of branches, the leaves be of a darke greene, somewhat long and broade, not muche unlike the leaves of Anagallis or Pimpernell, always two leaves growing one against an other, like the leaves of Anagallis. There groweth at the top of the stalke alongst the branches certayne floures one above an other, somwhat long and broade before, after the fasshion of a frogs mouth, not muche unlike the floures of Todeflaxe, but muche larger, and without tayles, of a faint yellowissh colour. After them comme long round huskes, the foremost part whereof are somewhat like to a Calfes snowte or Moosell, wherin the seede is conteyned."

In his *Pemptades* of 1583, Dodoens gives a glossary of botanical terms. His definitions suffer, however, from vagueness, and are not calculated greatly to advance the accurate description of plants. As an example we may take his account of the flower, which may be translated as follows:

"The flower (ἄνθος) we call the joy of trees and plants. It is the hope of fruits to come, for every growing thing, according to its nature, produces offspring and fruit after the flower. But flowers have their own special parts."

The descriptions from the pen of de l'Écluse are characterised by greater realism and closer attention to flower structure than those of his contemporaries. The plant which is now called *Sempervivum arboreum* L., of which his woodcut is reproduced in fig. 66 (below), is described as being "a shrub rather than a herb; occasionally it reaches the height of two Sedum maius.

Fig. 66. "Sedum majus", *Sempervivum arboreum* L. [de l'Écluse, *Rariorum...per Hispanias*, 1576]

cubits [3 ft.] and is as thick as the human arm, with a quantity of twigs as thick as a man's thumb: these spread out into numerous rays of the thickness of a finger. The ends of these terminate in a kind of circle, which is formed by numerous leaves pressing inwards all together and overlapping, just as in *Sedum vulgare majus*. These leaves however are fat and full of juice, and shaped like a tongue, and slightly serrated round the edge, with a somewhat astringent flavour; the whole shrub is coated with a thick, fleshy, sappy bark. The outer membrane inclines to a dark colour, and is speckled as in *Tithymalus characia*: the speckles are simply the remains of leaves which have fallen off. Meanwhile a thick pedicel covered with leaves springs out from the top of the larger branches, and bears, so to speak, a thyrsus of many yellow flowers, scattered about like stars, pleasant to behold. And when the flowers begin to ripen, and are running to seed (the seed is very small), the pedicel grows slender. But the plant is an evergreen."

In Gerard's *Herball* of 1597, the descriptions are seldom sufficiently original to be of much interest. We may quote, however, his account of the potato flower (fig. 67, p. 158), then so great a novelty that in his portrait he holds a spray of it in his hand (pl. xiv, f.p. 130). It will be noticed that he evidently paid greater attention to colour than to features that now appear to us to be more significant. The potato bears, he says, "very faire and pleasant flowers, made of one entire whole leafe, which is folded or plaited in such strange sort, that it seemeth to be a flower made of sixe sundrie small leaves, which cannot be easily perceived, except the same be pulled open. The colour whereof it is hard to expresse. The whole flower is of a light purple color, stripped down the middle of every folde or welt, with a light shew of yellownes, as though purple and yellow were mixed togither: in the middle of the flower thrusteth foorth a thicke fat pointell, yellow as golde, with a small sharpe greene pricke or point in the middest thereof."

The descriptions in the *Historia* of Valerius Cordus, published after his death, are among the best which were

written in the sixteenth century. Some of them are not merely
static descriptions of the species at maturity, but may include

Fig. 67. "Battata Virginiana", *Solanum tuberosum* L., Potato
[Gerard, *The Herball*, 1597]

an impression of the life history. Sometimes a chemical
interest is added by a careful indication of the taste of the
plant. We will not attempt to quote from Cordus, since

justice could not be done to his phytography without citations which would be too long for insertion here.

So far as the later herbals are concerned, the zenith of plant

Beta Cretica femine aculeato.

Fig. 68. "Beta Cretica semine aculeato"
[Bauhin, *Prodromos*, 1620]

description may be said to be reached in the *Prodromos* of Gaspard Bauhin (1620), in which a high level of terseness and accuracy is attained. As an example we may translate his description of " *Beta Cretica semine aculeato*", of which his drawing is reproduced in fig. 68, p. 159: "From a short tapering root, by no means fibrous, spring several stalks about 18 inches long: they straggle over the ground, and are cylindrical in shape and furrowed, becoming gradually white near the root with a slight coating of down, and spreading out into little sprays. The plant has but few leaves, similar to those of *Beta nigra*, except that they are smaller, and supplied with long petioles. The flowers are small, and of a greenish yellow. The fruits one can see growing in large numbers close by the root, and from that point they spread along the stalk, at almost every leaf. They are rough and tubercled and separate into three reflexed points. In their cavity, one grain of the shape of an *Adonis* seed is contained; it is slightly rounded and ends in a point, and is covered with a double layer of reddish membrane, the inner one enclosing a white, farinaceous core."

Any great advance upon Bauhin's descriptions could hardly be expected during the period discussed in the present book, since this period closed before the nature of the essential parts of the flower was understood. It was not until 1682 that the fact that the stamens are male organs was pointed out in print by Nehemiah Grew, though he himself attributed this discovery to Sir Thomas Millington, who is otherwise unknown as a botanist. Gerard's account of the stamens and stigma of the potato as a "pointell, yellow as golde, with a small sharpe greene pricke or point in the middest thereof", vague as it seems to the twentieth-century botanist, is by no means to be despised, when we remember that the writer was handicapped by complete ignorance of the functions and relations of the structures which he saw before him.

A marked feature of the descriptions written by the earlier herbalists was their reliance on *comparison*, in order to convey their notion of what a plant was like. In the description of the butterbur, copied from Bock, which we quoted from Fuchs'

2 *Plantago quinqueneruia rofea.*
Rofe Ribwoorte.

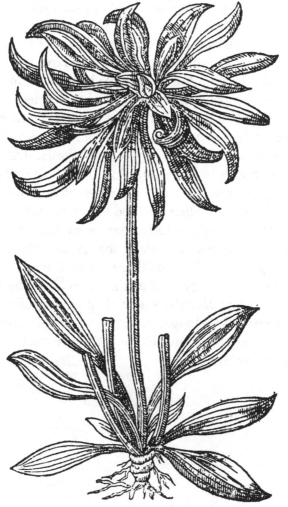

Fig. 69. "Rose Ribwoorte", an abnormal Plantain
[Gerard, *The Herball*, 1597]

V. Plant Description

herbal (p. 153), vine flowers, coltsfoot leaves, a mushroom, ashes, and a hat, are invoked as terms of similitude, and a small round table is used as an index of size. This amateur method was the only resource so long as the possibility of measurement was disregarded, and there was no terminology by means of which shape and arrangement could be expressed by reference to an agreed botanical standard. As we have already noticed, both Fuchs and Dodoens attempted glossaries, but such glossaries could not be constructed effectually in the state of knowledge of the time. It is a common lament at the present day, that botany has become incomprehensible, owing to the excessive use of technical words. There is, no doubt, some justice in this complaint, but, on the other hand, a study of the writings of the earlier herbalists makes it clear that a description of a plant couched in ordinary language—in which the meaning of the terms has never been defined in a botanical sense—generally fails at the critical points. It is to Joachim Jung and to Linnaeus that we owe the foundations of the accurate terminology now at the disposal of the botanist when he sets out to describe a new plant. The published work of these two writers belongs, however, to the late seventeenth and to the eighteenth centuries, and is thus outside the scope of the present volume.

Chapter VI

THE EVOLUTION OF PLANT CLASSIFICATION

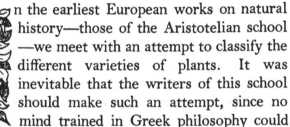n the earliest European works on natural history—those of the Aristotelian school —we meet with an attempt to classify the different varieties of plants. It was inevitable that the writers of this school should make such an attempt, since no mind trained in Greek philosophy could be content to leave a science in the condition of a mere chaos of isolated descriptions. Theophrastus, in the *Enquiry into Plants*, considers the principles of classification in a discussion which is still of interest. He suggests that the vegetable kingdom should be classed into trees, shrubs, under-shrubs, and herbs, and that minor divisions should be based upon such distinctions as those between cultivated and wild, flowering and flowerless, and deciduous and evergreen plants. He also hints at an ecological classification, but he makes it clear that all his proposals are of a tentative character.

Albertus Magnus, who kept alive in the middle ages the spirit of Aristotelian botany, followed the Theophrastean scheme in the main, but a study of his writings reveals the fact that he had in mind—though he never actually formulated it—a more highly evolved system, which may be represented, diagrammatically, as follows. The modern equivalents of his different groups are shown in square brackets, but it must not be assumed that he recognised the distinction between monocotyledons and dicotyledons as fully as this table may seem to imply:

I. Leafless plants [cryptogams in part].
II. Leafy plants [phanerogams and certain cryptogams].
   1. Corticate plants [monocotyledons].
   2. Tunicate plants ("ex ligneis tunicis") [dicotyledons].
     (*a*) Herbaceous.
     (*b*) Woody.

Cesalpino, whom we may treat as the typical Aristotelian botanist of the sixteenth century, makes his main distinction, on the old plan, between trees and shrubs, on the one hand, and undershrubs and herbs, on the other. He divides the first of these groups into two, and the second into thirteen classes, depending chiefly on what we should now call seed and fruit characters. Very few of these classes represent natural groups, but the recognition of the importance of the "seed" was a definite advance.

When we leave the botanical philosophers, and consider those who approached the subject from the standpoint of medicine, we find an altogether different state of affairs. The Aristotelian botanists were aware, from the beginning, of the philosophic necessity for some form of classification. The medical botanists, on the other hand, were interested in plants only as individuals, and were driven to classify them simply because some sort of order was necessary for convenience in dealing with a large number of kinds. In the *De materia medica* of Dioscorides, there is little attempt at arrangement, even in those versions which are not merely alphabetic. Book III, for instance, gives an account of roots, juices, herbs, and seeds, used for food or medicine, and the other books are defined with equal vagueness. Here and there, however, a slight feeling for kinship emerges; a number of umbellifers, for example, are enumerated in succession.

Pliny was not, strictly speaking, a medical botanist, but at the same time he may be mentioned in this connection, since his interest in plants was essentially utilitarian. Like Theophrastus, he begins his account of vegetation with the trees, but his reason for doing so is profoundly different from

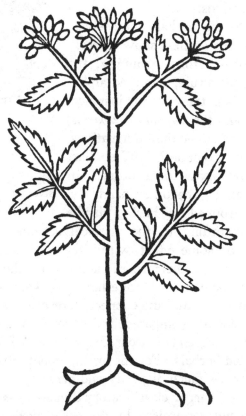

Fig. 70. "Carui" [*Ortus sanitatis,* 1491]

that of the Greek thinker, and illustrates the divergence between the scientific and the anthropocentric outlook upon the plant world. Theophrastus placed trees at the head of the vegetable kingdom, because he considered their organisation to be the most completely expressive of plant nature; Pliny, on the other hand, began with trees, because of their great

165

value and importance to man. As an example of his ideas of arrangement, we may mention that he sets the myrtle and the laurel side by side, on the ground that the laurel takes a corresponding place in triumphs to that accorded to the myrtle in ovations.

Turning to the herbals themselves, we find that the earliest show scarcely any trace of a natural grouping, the plants being, as a rule, arranged alphabetically. This is the case, for instance, in the *Latin* and the *German Herbarius*, the *Ortus sanitatis*, and their derivatives, and even in the great herbal of Leonhart Fuchs in the sixteenth century. In Bock's herbal, on the other hand, the plants are grouped as herbs, shrubs

Fig. 71. "Zparagus sive Asparagus" [*Ortus sanitatis, ?* 1497]

and trees, according to the classical scheme. The author evidently made some effort, within these classes, to arrange them according to their relationships. In the preface he writes:[1] "I have placed together, yet kept distinct, all plants which are related and connected, or otherwise resemble one another and are compared, and have given up the former old rule or arrangement according to the A.B.C. which is seen in the old herbals. For the arrangement of plants by the A.B.C. occasions much disparity and error."

Although the larger classificatory divisions, as now understood, were not recognised by the early workers, they had at least a dim understanding of the distinction between genera and species. Theophrastus, for instance, showed, by grouping together different species of oaks, willows, etc., that he had some conception of a genus. We owe to Konrad Gesner the first formulation of the idea that genera should be denoted by substantive names. He was probably the

[1] *Kreuter Bûch*, 1551.

166

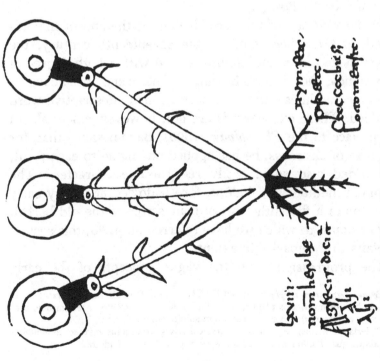

NOMEN HERBAE NYMPHEA.

A græcis dicit Prothea. Alii Caccabus. Alii Lo tometra. Alii Androcanos. Alii Hidrogogos. Alii Heracleos. Alii Arneon. Itali Nvmpheá.

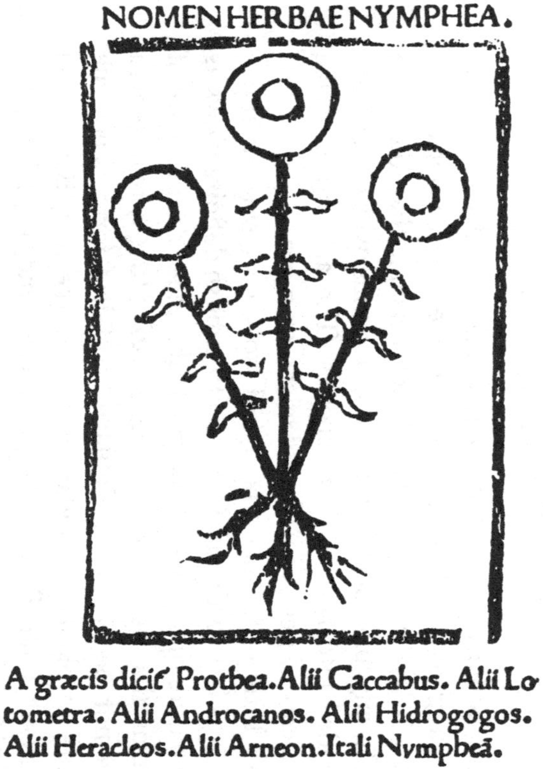

Fig. 73. "Nymphea", Water-lily [Printed version, *Herbarium Apuleii Platonici*, ?1481, after Hunger, F. W. T. (1935²)] *Reduced*

nymphæ.
ρboίλεc.
Caccocbiſ
coromeāſtæ

ḻ uίị
nom hæulg
Aλgẖ.ιτ δcίā
λ.χ²

Fig. 72. "Nymfea", Water-lily [Ninth-century manuscript of *Herbarium Apuleii Platonici*, *Codex Casinensis 97*, after Hunger, F. W. T. (1935²)] *Reduced*

167

earliest botanist who clearly expounded the distinction between a genus and a species. In one of his letters he writes: "And we may hold this for certain, that there are scarcely any plants that constitute a genus which may not be divided into two or more species. The ancients describe one species of gentian; I know of ten or more."

Very little of Gesner's botanical work ever saw the light, and it was left to Fabius Columna to publish definite views as to the nature of genera. It was no doubt under the influence of Cesalpino that he stated in his *Ekphrasis* (1616), that genera should not be based on similarities of leaf form, since the affinities of plants are indicated not by the leaf, but by the characters of the flower, the receptacle, and, especially, the seed.[1] He brought forward instances to show that previous authors had sometimes placed a plant in the wrong genus, because they attended only to the leaves and ignored the structure of the flower.

In the writings of Gaspard Bauhin, at the end of the sixteenth and the beginning of the seventeenth century, the binary system of nomenclature is used with a high degree of consistency, each species bearing a generic and specific name, though sometimes a third, or even a fourth, descriptive word is added. These extra words are not, however, essential. In the preface to the *Phytopinax* (1596), Bauhin states that, for the sake of clearness, he has applied one name to each plant, and added also some easily recognisable character.[2] The binomial method was, indeed, foreshadowed at a very early date, for in a fifteenth-century manuscript of the old herbal, *Circa instans*, to which we have referred on p. 26, this system prevails to a remarkable extent.

The progression from the vague concepts of the early

---

[1] *Minus cognitarum stirpium*...ΕΚΦΡΑΣΙC. 1616. Pars altera, Cap. xxvii. p. 62 "tam in hac, quam in aliis plantis, non enim ex foliis, sed ex flore, seminisque, conceptaculo, et ipso potius semine, plantarum affinitatem dijudicamus."

[2] "plerisque nomen imposuimus, perspicuitatis gratia, cuius nomine communiter nota aliqua quae à quolibet in planta observari potest, nomini addita."

writers to the sharp definition of genera and species to which we are now accustomed, has been in some ways a doubtful blessing. There is to-day, as a recent writer has pointed out, a tendency to treat these units as if they possessed concrete reality, whereas they are merely convenient abstractions,

NENVFAR

Fig. 74. "Nenufar", Water-lily [*Latin Herbarius* (Arnaldus de Villa Nova, *Tractatus de virtutibus herbarum*), 1499]

which make it easier for the human mind to cope with the endless multiplicity of living things.

When we turn from the consideration of genera and species, to the wider systems of classification which were evolved by the earlier herbalists, we are at once struck by the marked difference between the *principles* upon which these schemes are based, and those at which we have arrived at the

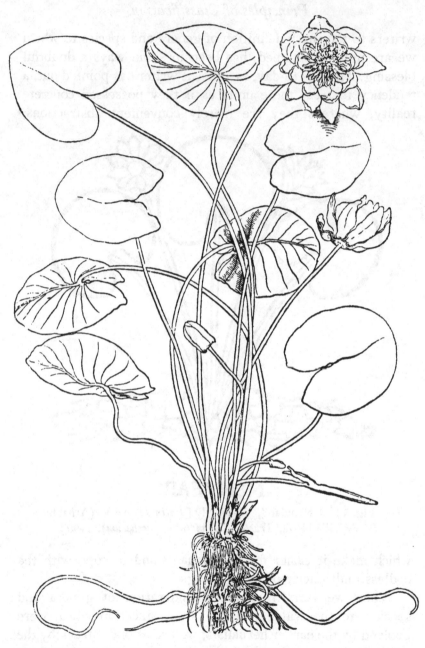

Fig. 75. "Nenuphar", *Nymphaea alba* L., White-water-lily [Brunfels, *Herbarum vivae eicones*, vol. I, 1530] *Reduced*

present day. To classify plants according to their uses and medicinal properties is obviously the first suggestion that arises, when the universe is regarded from the crudely anthropocentric standpoint. In *The grete herball* of 1526 we get a ludicrously clear example of this method, applied to the special case of the fungi. "Fungi ben mussherons.... There be two maners of them, one maner is deedly and sleet [slayeth] them that eateth of them and be called tode stoles, and the other dooth not." The same account of the fungi occurs also in the earlier herbal, *Circa instans*. Although it seems at first glance completely unscientific, it must be admitted that this theory of classification according to "virtues" has within it the germ of something approaching to a natural system. Both Linnaeus and de Jussieu have pointed out that related plants have similar properties, and, in 1804, A. P. de Candolle, in his *Essai sur les propriétés médicales des Plantes, comparées avec leurs formes extérieures et leur classification naturelle*, carried the argument much further. He showed that, in no less than twenty-one families of flowering plants, the same medicinal properties were found throughout all the members. This is the more remarkable when we realise that de Candolle was obliged to disregard a large number of families, since their properties were then unknown.

In the history of botanical classification, the first advance from the purely utilitarian standpoint was marked by the recognition of the fact that the structure and mode of life of the plants themselves, as well as their uses, might be treated as significant. We have pointed out that, in the *Historia plantarum Lugdunensis* of d'Aléchamps (1586,7), much of the knowledge of the period was brought together, so we may take it as a typical example. In this herbal we find the issues curiously confused by the unrelated working of three different principles; that is to say, by the simultaneous insistence (i) on the habitat, (ii) on the "virtues", and (iii) on the

171

structure, of the plant, as affording the prime clue to its systematic position. The herbalist thus erects his scheme on

Fig. 76. "Gele Plompen", *Nuphar luteum* Sm., Yellow-water-lily
[de l'Obel, *Kruydtboeck*, 1581]

the basis of an undigested medley of ecological, medical, and morphological ideas. An enumeration of the eighteen headings, under which d'Aléchamps described the vegetable

kingdom, so far as it was then known, will show the per-
plexities which surrounded the first gropings after a natural
system. His headings may be translated as follows:

I. Of trees which grow wild in woods.
II. Of fruits growing wild in thickets and shrubberies.
III. Of trees which are cultivated in pleasure gardens and
orchards.
IV. Of cereals and pulse, and the plants which grow in the
field with them.
V. Of garden herbs and pot herbs.
VI. Of umbelliferous plants.
VII. Of plants with beautiful flowers.
VIII. Of fragrant plants.
IX. Of plants growing in marshes.
X. Of plants growing in rough, rocky, sandy, and sunny
places.
XI. Of plants growing in shady, wet, marshy, and fertile
places.
XII. Of plants growing by the sea, and in the sea itself.
XIII. Of climbing plants.
XIV. Of thistles and all spiny and prickly plants.
XV. Of plants with bulbs, and succulent and knotty roots.
XVI. Of cathartic plants.
XVII. Of poisonous plants.
XVIII. Of foreign plants.

Among these eighteen groups, the only ones which have
any pretension to being natural are VI (umbellifers) and
XIV (thistles), and these merely approximate roughly to
related groups of genera. Among the umbellifers, we meet
with *Achillea* and other genera which do not really belong to
the family, whilst, with the thistles, there are grouped some
spiny plants, such as *Astragalus Tragacantha* L., which, in a
natural system, would occupy a place remote from the com-
posites.

VI. Plant Classification

More than fifty years after the appearance of d'Aléchamps' herbal, John Parkinson, in his *Theatrum botanicum* of 1640, propounded another classification of a most rudimentary type. He divided all the plants then known into seventeen classes or tribes—the sequence in which these classes were placed having, in most cases, no meaning at all. A few of his

Fig. 77. "Ninfea", Water-lilies [Durante, *Herbario Nuovo*, 1585 (after Mattioli)]

tribes are natural, but many are valueless as expressions of affinity. As an example we may mention his third class, "Venemous, Sleepy, and Hurtfull Plants, and their Counter-poysons," and his seventeenth, "Strange and Outlandish Plants." In Parkinson's classification, we see botany reverting once more to the position of a mere handmaid to medicine.

For the evolution of better ideas about plant classification, we have to go back to the sixteenth century. When we consider the botanists of the Low Countries, we find that de l'Écluse took no special interest in the subject, but that Dodoens made an attempt at a general scheme, though without any adequate guiding principle. Within the larger divisions, he showed, however, some perception of natural affinities. He often assembled genera which we now regard

174

ASPARAGVS

Spargen.

Fig. 78. "Asparagus", *A. officinalis* L. [Fuchs, *De historia stirpium*, 1542] *Reduced*

175

as members of the same family, and species which we now look upon as belonging to a single genus. He brought together, for instance, certain plants which are members respectively of the Geraniaceae, Hypericaceae, Plantaginaceae, Compositae, etc. His recognition of some families was, however, markedly imperfect; among the Umbelliferae, for example, he described love-in-a-mist (*Nigella*) and a couple of saxifrages. Such vagaries show how little stress was laid upon the flowers and fruits at this time, from the point of view of classification. Moreover, though much importance was attributed to the shape of the leaf, there was no clear insight into the meaning of its variations. It was left to Dodoens' fellow-countryman, de l'Obel, to observe leaf form with a seeing eye, and to work out a classification, based mainly on foliar characters, which was a considerable advance on previous efforts. He put forward his system in his *Stirpium adversaria nova* (1570,1), and used it also in his later work. It was thus published much earlier than the very primitive schemes of d'Aléchamps and Parkinson, to which we have referred above. The best point of his system is that, by reason of their characteristic differences of leaf structure, he distinguishes the classes now known to us as monocotyledons and dicotyledons. He also introduces a useful feature in the shape of a synoptic table of species which precedes each more or less natural group of plants. The superiority of his classification to the other contemporary arrangements was immediately realised. There is evidence of this in the fact that, after de l'Obel's *Kruydtboeck* was published, Plantin, as we have already mentioned, brought out an album of the wood-engravings used in the book, which, although they had also appeared as illustrations to the works of Dodoens and de l'Écluse, were now arranged as in the scheme set forth by de l'Obel," according to their kind and their mutual relationship."[1]

---

[1] "uti à D. Mathia Lobelio...singulae videlicet congeneres ac sibi mutuo affines digestae sunt." Dedication to *Plantarum seu stirpium icones*, 1581.

De l'Obel's Classification

There seems little doubt that de l'Obel made a more conscious effort than any of his predecessors to arrive at a natural classification, and that he felt that such a classification would reveal an underlying unity in all forms of life. In the preface to his *Stirpium adversaria nova* of 1570,1, he wrote of an *ordo universalis*, in which "things which are far and widely different become, as it were, one thing".

De l'Obel's scheme is not expressed in the clear manner to which we have become accustomed in more modern systems, because, in common with other botanists of his time, he did not, as a rule, give names to the groups which we now call families, or draw any sharp line of distinction between them. His arrangement, in spite of its good features, had serious drawbacks. The anomalous monocotyledons, such as lords-and-ladies (*Arum*), black-bryony (*Tamus*), and butchers'-broom (*Ruscus*), are scattered among the dicotyledons, while the sundew (*Drosera*) appears with the ferns, and so on. Similarities of leaf form, which are now regarded merely as instances of parallelism in evolution, are responsible, in his system, for many meaningless groupings. For instance in the *Kruydtboeck* we find the twayblade (*Listera*), the may-lily (*Maianthemum*) and the plantain (*Plantago*) described in succession, while, in another part of the book, various clovers (*Trifolium*), wood-sorrel (*Oxalis*), and *Anemone Hepatica* L., are grouped together. It is less unreasonable that the marsh-marigold (*Caltha*), the water-lilies (*Nymphaea* and *Nuphar*), the fringed-water-lily, *Limnanthemum*, and frogbit (*Hydrocharis*) should follow one another, or that de l'Obel should have brought together the broomrape (*Orobanche*), the toothwort (*Lathraea*), the bird's-nest-orchid (*Neottia*) and a number of fungi. In these latter instances he has arrived, in fact, at genuine biological (though not morphological) groupings. He has recognised, on the one hand, the marked uniformity in the type of leaf characteristic of "swimming" water-plants, and, on the other hand, he has observed the

Fig. 79. "Tussilago", *T. Farfara* L., Coltsfoot [Fuchs, *De historia stirpium*, 1542] *Reduced*

leaflessness, and the absence of green colour, which are negative features common to so many saprophytes and parasites.

The perception of natural affinities among plants which, in the sixteenth and seventeenth centuries, was gradually, in a dim, instinctive fashion, arising in men's minds, is perhaps best expressed in the writings of Gaspard Bauhin, especially in his *Pinax theatri botanici* (1623). This work is divided into twelve books, each book being further sub-divided into sections, comprehending a variable number of genera. Neither the books nor the sections have, as a rule, any general heading, but there are certain exceptions. For instance, Book II is called "de Bulbosis", and a section of Book IV, including eighteen genera, is headed "Umbelliferae". Some of the sections represent truly natural groups. Book III, Section VI, for example, consists of ten genera of Compositae, while Book III, Section II includes seven Cruciferae. Other sections contain plants of more than one family, but yet show a distinct feeling for relationship. For instance, Book V, Section I includes *Solanum, Mandragora, Hyoscyamus, Nicotiana, Papaver, Hypecoum* and *Argemone*—that is to say, four genera from the Solanaceae followed by three from the Papaveraceae. The common character which brings them together here is, no doubt, their narcotic property, but, although no definite line was drawn between the plants belonging to these two widely sundered families, the order in which they are described shows that their distinctness was recognised. Some of Bauhin's other groups, however, which, like that just discussed, are distinguished by their properties, or, in other words, by their chemical features, have no pretension to naturalness from a morphological standpoint. This is the case with the group described in Book XI, Section III under the name of "Aromata", which consists of a heterogeneous assemblage of genera belonging to different families, which are only connected by the fact that they all yield spices useful to man.

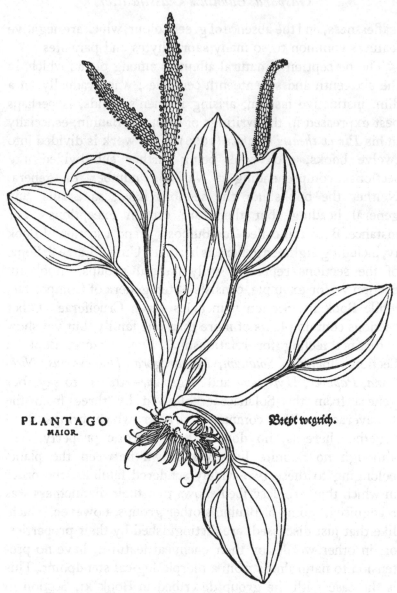

PLANTAGO
MAIOR.

Breyt wegrich.

Fig. 80. "Plantago major", Plantain [Fuchs, *De historia
stirpium*, 1542] *Reduced*

180

Gaspard Bauhin and Adam Zaluziansky

There is no doubt that, on the whole, Bauhin was markedly successful in recognising affinities within small cycles, but he broke down on the broader question of the relationships between the groups of genera so constituted. This is, however, hardly surprising when we remember how much difference of opinion exists among systematic botanists, even to-day, upon the subject of the relations of the families to one another.

Like de l'Obel, Bauhin seems to have believed in the general principle of a progression from simpler to more highly developed forms. His application of this principle led him to begin with the grasses and to conclude with the trees. The question as to which groups among the flowering plants [angiosperms] are to be considered as relatively primitive, is still, at the present day, an open one, but it would be conceded generally that Bauhin's arrangement cannot be accepted. In their own way, the grasses are highly organised, while the "tree habit" seems to have been adopted independently by many plants belonging to widely differing cycles of affinity.

On the subject of the relation of the cryptogams (flowerless plants) to the flowering plants, Bauhin evidently had only indistinct ideas. We find, for instance, the ferns, mosses, corals (!), fungi, algae, the sundew, etc., sandwiched between some Leguminosae and a section consisting chiefly of thistles. Much clearer notions about the difference between the lower and higher forms of life had been expressed as early as 1592 by the Bohemian botanist, Adam Zaluziansky von Zaluzian. He pointed out that some animals, such as sponges, are "informia et indigesta" (without form or ordered structure), while others are "absolutiora" (of a higher degree of perfection), and among plants he recognised the same kind of range. Corresponding to the lower animals, there are forms of vegetable life that are "ruda et confusa" (formless and lacking in order); he places fungi in this class, and also seaweeds, lichens, and other plants, which he includes under the general term "musci". He then proceeds to the beings which he regards as more perfect (the flowering plants, etc.),

Fig. 81. "Althaea Thuringica", *Lavatera thuringiaca* L.
[Camerarius, *Hortus medicus,* 1588]

which he describes in a series beginning with the simple-leaved grasses and rushes. He places ferns in this higher series, associating them with other plants with leaves of elaborate form, but he notes that the ferns have a totally different mode of reproduction.

In the significance ascribed to the degree of elaboration of the leaf, Zaluziansky's views were in marked contrast with those published in the previous decade by Cesalpino. This

Fig. 82. "Pulsatilla", *Anemone Pulsatilla* L., Pasque-flower [Camerarius (Mattioli), *De plantis Epitome*, 1586]

Italian savant, as we have already mentioned, had arrived at the principle that seed and fruit characters were of supreme value in classification. The system which he deduced from this principle was, however, relatively unrealistic. The reason for its inadequacy may probably be sought in the fact that

VI. Plant Classification

no system of classification can be natural, unless it takes into account the nature of the plant as a whole. Indeed, the perfect classification, which even to-day remains a remote ideal, must be based not only upon the external features of the vegetative and reproductive systems, but also upon the anatomy, chemistry, and cytology of the plant, as well as upon its life history, genetic behaviour, and pathological reactions. Cesalpino failed, because he deliberately selected a narrow basis for his classification, since he relied primarily upon one organ, the "seed". It is perhaps due to the inhibiting effects of this restricted method, that he failed to grasp the fundamental distinction between the embryos of dicotyledons and monocotyledons, and thus did not even reap the obvious harvest of his own theory. Brilliant as Cesalpino's mind undoubtedly was, he followed the wrong path in treating the plant as a congeries of separate members, of which one category could be selected on *a priori* grounds as affording the clue to taxonomy. A profounder insight into natural affinity was the reward of those herbalists who approached the subject without a preconceived theory, and felt their way towards a classificatory system, with the aid of that synthetic common sense to which the plant is a *whole*, individualised and indivisible.

Chapter VII

THE EVOLUTION OF THE ART OF BOTANICAL ILLUSTRATION

n the art of botanical illustration, we do not find, in Europe, a steady advance from early pictures of poor quality to later ones of a finer character. On the contrary, among the earliest extant drawings of a definitely botanical intention, there are modern-looking figures, free from such features as would be now generally regarded as archaic. The large-scale brush drawings in the famous Vienna codex of Dioscorides (pp. 8, 9) are remarkable examples of the excellence of some of the very early work (pls. i, f.p. 4; ii, f.p. 10; xviii, f.p. 186; xxiii, f.p. 240). The date of this codex is about A.D. 512, but the pictures are of much older extraction, being probably derived, in part at least, from originals by Krateuas belonging to the first century before Christ. The general habit of the plant is often admirably expressed, and sometimes the characters of the flowers and seed vessels are well indicated. Even the best of these pictures, however, show signs of having suffered in a process of copying and recopying, extending over several centuries.

The study of the illustrations in the manuscript herbals belonging to the period of nearly a thousand years which intervened between the making of Anicia Juliana's copy of Dioscorides and the appearance of the first printed herbals, is a world in itself. We must limit ourselves to such slight incursions into this world as are absolutely necessary for the

185

VII. Botanical Illustration

understanding of the printed book, since our primary concern is with the evolution of botanical figures in the two centuries after the setting up of the first printing presses. We will now merely say—speaking in the most generalised terms—that the history of plant illustration in the manuscript era is a history of degradation rather than of progress.

For the earlier part of our special period, the significant fact was the discovery of the art of printing pictures from wood-blocks, by means of lines left raised by cutting away the intervening areas. Since there is still much variety of opinion amongst experts as to the difference, if any, between the meaning to be attached to the terms "woodcut" and "wood-engraving", it seems best to use these expressions in the present book as if they were synonymous.

Botanical illustration in the period of wood-engraving may be treated as belonging to two schools, but it should be understood that this classification is somewhat arbitrary. The first of these schools may be regarded as representing the last decadent expression of that tradition of late classical art, which, more than a thousand years earlier, had given rise to the drawings on which those in the Vienna manuscript were based. Probably no original woodcuts produced after the close of the fifteenth century should be assigned to this school. Copying was of its essence; it was, indeed, merely a belated continuation of the manuscript tradition into the era of the printed book. We shall not do justice to this phase unless we realise that, before there were any mechanical means of multiplying either the text or the figures, literal copying by hand possessed a virtue and a value which it has lost to-day. In the second school of botanical engraving, which culminated artistically, if not scientifically, in the sixteenth century, there was a marked change; the herbal artists cast off the fetters of the manuscript tradition, and returned to the direct observation of the living plant.

The figures of the first school, of which the cuts in the

Plate xviii

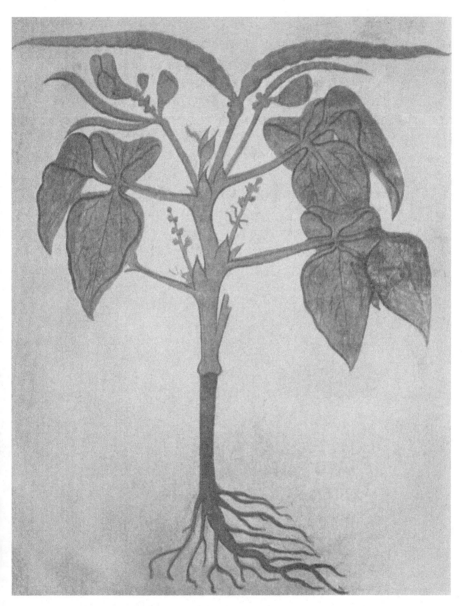

ΦΑCIΟΛOC, *Vigna unguiculata* (L.) Walp. (or *V. sinensis* Endl., *Dolichos ungui-
culatus* L. or *D. Lubia* Forsk.) [Dioscorides, *Codex Aniciae Julianae* (Vind. Med.
Gr. I), circa A.D. 512, facsimile, 370 verso] *Reduced*

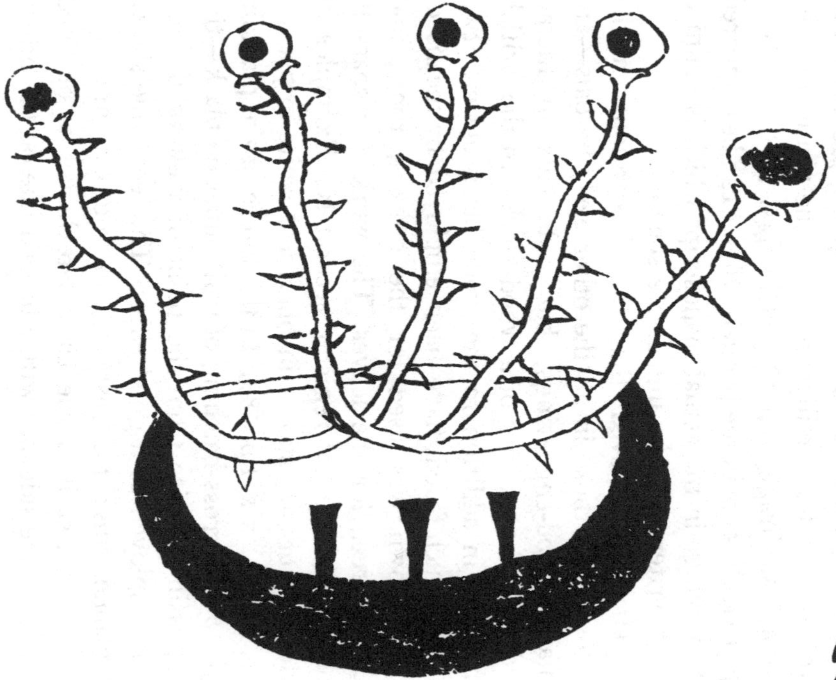

NOMEN HERBAE SAXIFRAGIA.

A grzcis dicitur
Alii

Adiantos.
Scolopendriæ.

Fig. 84. "Saxifragia" [Printed version, *Herbarium Apuleii Platonici*, ?1481, after Hunger, F. W. T. (1935²)] *Reduced*

Fig. 83. "Saxifraga" [Ninth-century manuscript *Herbarium Apuleii Platonici, Codex Casinensis 97*, after Hunger, F. W. T. (1935²)] *Reduced*

187

VII. Botanical Illustration

Roman edition of the *Herbarium* of Apuleius Platonicus (?1481) may be taken as examples, are formal and decorative. Derived from a long manuscript tradition, they have, to a great extent, lost touch with nature. Even in the first century of our era, Pliny lamented that the pictures in the illustrated herbals had become degraded in copying, especially in regard to the colouring. He pointed out (in the words of Philemon Holland's translation) that "they that limned and drew them out, did faile and degenerat from the first pattern and originall". If Pliny felt this about the herbals of his time, one wonders how he would have found language strong enough to condemn the figures in the printed *Herbarium*, which were the last term in a copying process which may have gone on for a millennium. In the pictures of the water-lily and saxifrage, reproduced in figs. 72, 73, p. 167; 83, 84, p. 187, the cuts in the printed volume can be compared with the drawings in the actual manuscript which is claimed by a recent writer to be its immediate source.

It is generally believed that the illustrations in the *Herbarium*—unlike those in all the other early herbals—are not actually wood-engravings, but rude cuts in metal, excavated after the manner of a wood-block. In the British Museum copy, in addition to the crude black outline, two colours, now much faded, have been employed by means of stencilling; brown was used for the flowers, roots, and animals, and green for the leaves. The work was coarsely done, and the tints do not register accurately. Examples of these coloured cuts are shown in black and white in pls. iv, f.p. 16; v, f.p. 18; xix, f.p. 188. In the figure of the plantain (fig. 1, p. 15) the cross-hatching of white lines on black—the simplest possible device from the point of view of the engraver—is employed with good effect. The technique shows various mannerisms; for instance, in the cut of "Vettonica" (Betony), each of the lanceolate leaves is outlined continuously on one side, but with a broken line on the other.

Plate xix

"Dracontea" [*Herbarium Apuleii Platonici,* ? 1481]
(*The tint represents contemporary colouring*)

Illustrations in Incunabula

The figures in the *Herbarium* are characterised by an excellent trait, which is common to most of the older herbals, namely the habit of portraying the plant as a whole, including its roots. No doubt this habit arose because the root was often of special value from the druggist's point of view. It is to be regretted that, in modern botanical drawings, the recognition of the paramount importance of the flower and fruit, in classification, has led to a comparative neglect of the vegetative organs, especially those which exist underground.

We next come to a series of pictures, which may be regarded as intermediate between the classical tradition of the *Herbarium* of Apuleius, and the renaissance of botanical drawing, which took place early in the sixteenth century. This series includes the illustrations to the *Book of Nature*, and to the *Latin Herbarius*, the *Herbarius zu Teutsch*, the *Ortus sanitatis*, and their derivatives, which were discussed in Chapters II and III.

Das pùch der natur of Konrad von Megenberg occupies a unique position in the history of botany, for it is the earliest work in which woodcuts representing plants were used with the definite intention of illustrating the text, and not merely for a decorative purpose. It was first printed in Augsburg in 1475, and is thus several years older than the earliest printed edition of the *Herbarium* of Apuleius Platonicus. The drawings which it contains are probably not of such great antiquity, however, as those of the *Herbarium*, for the appearance of the only plant illustration which survives in the British Museum example (pl. iii, f.p. 14), suggests that this picture had not had such a long career of copying and recopying before it was engraved. It shows a number of plants growing *in situ*, among which a buttercup, violet, and lily-of-the-valley are distinctly recognisable. It is noticeable that, in two cases in which a rosette of radical leaves is represented, the centre of the rosette is indicated in black, upon which the leaf stalks appear in white. This use of the black background, which gives a rich and solid effect, was carried much further in later books, such as the *Ortus sanitatis.*

Woodcuts, somewhat similar in style to that just described, but more primitive, occur in Trevisa's version of the mediaeval encyclopaedia of Bartholomaeus Anglicus, which was printed

Fig. 85. "Brionia" [*Latin Herbarius,* 1484]

by Wynkyn de Worde before the end of the fifteenth century. They were probably the first figures, of strictly botanical aim, illustrating an English book. One of them is reproduced in fig. 19, p. 42.

The illustrations to the *Latin Herbarius,* or *Herbarius Moguntinus,* published at Mainz in 1484 (figs. 3, p. 18; 4, p. 19; 5, p. 20; 85, p. 190), form the next group of botanical woodcuts. These figures are much better than those of the *Herbarium* of Apuleius, but at the same time they are, as a rule, formal and conventional, and often quite unrecognisable. The want of realism is conspicuous in such a drawing as that of the lily (fig. 3, p. 18), in which the leaves are represented as if they had no organic continuity with the stem.

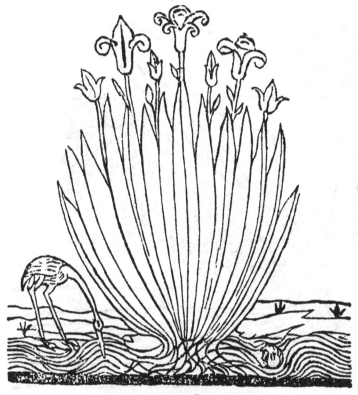

IREOS VEL IRIS

Fig. 86. "Ireos vel Iris", probably *Iris Pseudacorus* L., Yellow-flag [*Latin Herbarius* (Arnaldus de Villa Nova, *Tractatus de virtutibus herbarum*), 1499]

Some of the figures have a special charm, and, in their decorative effect, recall the plant designs so often used in the middle ages to enrich the borders of illuminated manuscripts. The bryony (fig. 85, p. 190), which might well have come from the edge of a missal, has the conventionalised form of tendril which is also employed in other early work—for instance, the roof-painting of a vine in the Chapel of St Andrew, Canterbury Cathedral.

Another series of figures, also illustrating the text of the *Latin Herbarius*, was published a little later, in Italy. The

Fig. 87. "Capillus Veneris", ? Maidenhair-fern [*Latin Herbarius* (Arnaldus de Villa Nova, *Tractatus de virtutibus herbarum*), 1499]

woodcuts are believed to be derived mostly from German sources. Figs. 6, p. 21; 63, p. 150; 74, p. 169; 86, p. 191; 87; 88, p. 194, are taken from a Venetian edition of 1499. These drawings are more ambitious than those in the original German issue, and, on the whole, they are more naturalistic. A delightful example, almost Japanese in style, shows an iris growing at the margin of a stream, from which a graceful bird is drinking (fig. 86, p. 191). In another picture (fig. 87, p. 192), the fern called "Capillus Veneris", which is perhaps intended for the maidenhair, is represented hanging from rocks over water. The very symmetrical drawing of the peony (fig. 63, p. 150) is remarkable for its attempt to represent the tuberous roots, which are indicated in solid black. In the no less symmetrical water-lily (fig. 74, p. 169), the scars of the leaf bases are emphasised diagrammatically. This drawing is of interest, also, on account of its frank disregard of proportion; the flower stalks are drawn not more than twice as long as the breadth of the leaf. This disregard of the scale of the parts is seen also in the aroid of the *Herbarium* of Apuleius Platonicus (pl. xix, f.p. 188). We may safely conclude that the draughtsman knew quite well, in such cases, that he was not representing the plant as it was, and that he intentionally gave a conventional rendering, which did not profess to be more than an indication of certain distinctive features. This attitude of the artist to his work as symbolism, which is so different from that of the scientific draughtsman of the present day, is seen with great clearness in many of the drawings in mediaeval manuscripts. For instance, a plant such as the houseleek may be represented growing on the roof of a house—the plant being about three times the size of the building. It is evident that the little house was introduced merely to convey graphic information as to the habitat of the plant concerned, and that the scale on which it was depicted was simply a matter of convenience. Before an art can be appreciated, its conventions must be accepted. It would be as absurd to quarrel with the illustrations we have just described, on account of their lack of proportion, as to condemn grand opera because, in real life, men and women do not converse in song.

In 1485, the year following the first appearance of the *Latin Herbarius*, the very important work known as the *German Herbarius*, or *Herbarius zu Teutsch*, made its appearance at Mainz. As we noticed in Chapter II, its illustrations—three of which, the iris, the dodder, and the winter-cherry, are shown in figs. 7, p. 27; 89, p. 195; and 90, p. 196—are often of remarkable beauty. Their greater realism, as compared with those in the *Latin Herbarius,* is no doubt partly

CVSCVTA

Fig. 88. "Cuscuta", Dodder [*Latin Herbarius* (Arnaldus de Villa Nova, *Tractatus de virtutibus herbarum*), 1499]

due to their larger size. Naturalism in the earlier botanical woodcuts tended, indeed, to be a function of scale. The technique was not delicate enough to give good scientific results when the blocks were small.

A pirated second edition of the *Herbarius zu Teutsch* appeared at Augsburg only a few months after the publica-

194

Fig. 89. "Cuscuta", Dodder [*Herbarius
zu Teutsch*, Mainz, 1485]

Fig. 90. "Alkekengi", *Physalis*, Winter-cherry [*Herbarius zu Teutsch*, Mainz, 1485]

tion of the first at Mainz. The figures, which are roughly copied from those of the original edition, are very inferior to them. In fact, the Mainz woodcuts of 1485 excel those of all subsequent issues.

In the *Ortus sanitatis* of 1491, about two-thirds of the

Fig. 91. "Alkekengi", *Physalis*, Winter-cherry
[*Ortus sanitatis*, 1491]

drawings of plants are derived from the *Herbarius zu Teutsch*. They are often much spoiled in the process, and it is evident that the copyist frequently failed to grasp the intention of the original artist. The woodcut of the dodder (fig. 92, p. 198) for instance, is lamentably inferior to that in the *Herbarius zu Teutsch* (fig. 89, p. 195). The process of debasement in copying

197

seems to depend upon certain definite psychological trends, some of which are revealed on comparing, for example, the figures of the winter-cherry in the *Herbarius zu Teutsch* (fig. 90, p. 196) and in the *Ortus sanitatis* (fig. 91, p. 197). One of these trends is the hankering after symmetry, arising probably

Fig. 92. "Cuscuta", Dodder [*Ortus sanitatis*, 1491]

from the fact that human beings are themselves more or less symmetric objects. This instinct emerges in the version of the winter-cherry in the *Ortus sanitatis*. Here the copyist disregarded the axillary nature of the right-hand shoot, because it produced an asymmetric effect which offended his eye; to avoid this asymmetry, he added a left-hand leaf, which did not exist in the original, and bent the right-hand branch to

198

give the effect of a dichotomy, thus destroying the character of the shoot system. The group of leaves at the top was not well realised in the original, so, to save himself trouble, the copyist substituted a thorn-like process for one of them. He then apparently felt that the space between the two branches

Fig. 93. "Botris" [*Ortus sanitatis*, 1491]

was unsatisfactory, and added a similar thorn to mitigate the blankness. Finally, regarding the simple cut end of the branch as too lame a conclusion, he finished it off with a writing-master's flourish. Such flourishes, which are recurrent in the *Ortus*, illustrate the dangers of dexterity. The craftsman, who cut the blocks for the *Herbarius zu Teutsch*, could scarcely have produced these curlicues, so he was saved

199

by his lack of technical finesse from the pretentiousness to which the engraver of the *Ortus sanitatis* was prone.

The use of a black background, with the stalks and leaves forming a contrast in white, which we noticed in *The Book of Nature*, is carried further in the *Ortus sanitatis*. This device is employed with effect in the tree-of-knowledge (fig. 14, p. 33),

Capitulū.cccclꝛꝛiij

Fig. 94. "Tilia", Lime-tree with birds
[*Ortus sanitatis*, ? 1497]

and in fig. 10, p. 30; fig. 93, p. 199; and fig. 94. No consistent method is followed in the coarse shading which is introduced. In some cases there seems to have been an attempt at the convention, used so successfully by the Japanese, of darkening the underside of the leaf, but, sometimes, in the same figure, certain leaves are treated in this way, and others not. In some of the genre pictures, Noah's-ark trees are introduced, with crowns consisting entirely of parallel horizontal lines,

200

decreasing in length from below upwards, so as to give a triangular form.

The picture of a lime-tree, in which birds of unusual aspect are perching (fig. 94, p. 200), illustrates the deliberate disregard of proportion to which we have already referred as characterising the manuscript and incunabula herbals. By ignoring questions of relative scale, the artist is enabled to indicate the fact that grass can grow beneath lime-trees, and to show the type of trunk, and also the shape of the leaf—points which could not have been made clear if the natural proportions had been respected.

The grete herball and a number of other works of the early sixteenth century derived from the *Herbarius zu Teutsch*, the *Ortus sanitatis*, and similar sources, are of no importance in the history of botanical illustration, since scarcely any of their figures are original. The oft-repeated set of woodcuts, based upon those in the *Herbarius zu Teutsch*, were also used to illustrate Hieronymus Braunschweig's treatise (1500), translated in 1527 as *The boke of Distyllacyon*. That the conventional figures of the period were generally regarded as inadequate, may be deduced from some remarks by Hieronymus at the conclusion of his work. He tells the reader that he must attend to the text rather than the figures, "for the figures are nothing more than a feast for the eyes, and for the information of those who cannot read or write".[1] That these figures could, indeed, hardly have satisfied the demands of the time, is suggested, for instance, by the existence of such naturalistic work as the carvings of foliage, flowers, and fruit, on the capitals of the Southwell Chapter House, which are more than two hundred years earlier than *The grete herball* and *The boke of Distyllacyon*.

During the first three decades of the sixteenth century, the art of botanical illustration thus remained in abeyance in

[1] "wan die figuren nit anders synd dann ein ougenweid und ein anzeigung...die weder schriben noch lesen kündent."

VII. Botanical Illustration

Europe. Such books as were published were supplied, as we have seen, with mere copies of older woodcuts. In 1530, however, a new era was inaugurated with the appearance of Brunfels' great work, the *Herbarum vivae eicones*, in which a number of plants native to Germany, or commonly cultivated there, were drawn with a beauty and a fidelity which have rarely been surpassed (figs. 22, p. 53; 23, p. 54; 24, p. 56; 25, p. 57; 75, p. 170; 97, p. 205; 98, p. 207; 99, p. 208). There is possibly some significance in the fact that a date round about 1530 has been taken as the limit of the "Gothic" period and the beginning of the renaissance by the students of another art—stained glass.

Brunfels' illustrations represent so notable an advance on any previous botanical woodcuts, that, before considering them in detail, it will be well to seek for the causes of this sudden improvement. On taking a general view of the subject, we find that, at the beginning of the sixteenth century, there was a forward movement in all the branches of book illustration, and not merely in the one branch with which we are here concerned. The impetus to this movement seems to have come from the fact than many of the best artists, such as Albrecht Dürer, began at that period to draw for wood-engraving, whereas in the fifteenth century the ablest men had shown a tendency to despise the craft and to hold aloof from it. The engravings in Brunfels' herbal, and the beautiful books which succeeded it, should not, then, be treated as if they were an isolated manifestation, but should be viewed in relation to other contemporary or even earlier plant pictures, which were not intended for book illustration. An exquisite drawing by Leonardo da Vinci, whose career closed eleven years before the appearance of the *Herbarum vivae eicones*, is reproduced in pl. xx. It is a surprising thought that numerous editions of the *Ortus sanitatis*, and similar works, with their rough and crude woodcuts, were published during the years when Leonardo da Vinci was at

Plate xx

Study of *Ornithogalum umbellatum* L., Star-of-Bethlehem, and other plants [Leonardo da Vinci, 1452–1519. Drawing in the Royal Library, Windsor] *Reduced*

the summit of his powers. If internal evidence alone were available, it might be argued, with some plausibility, that the engravings in the *Ortus sanitatis*, and the drawings of Leonardo da Vinci, were centuries apart—and so, perhaps, they were, if, as is possible, the printed sources from which the *Ortus sanitatis* figures were derived, had a long manuscript tradition behind them.

Albrecht Dürer, also, produced remarkable flower paintings

De Ranis.

Fig. 95. Aquatic animals and marsh plants [F. Boussuet, *De Natura aquatilium*. Lugduni, apud Matthiam Bonhome, 1558]

in the period before Brunfels' herbal. His picture of the columbine (pl. xxi, f.p. 204), drawn in 1526, will serve as an example. The work of Leonardo da Vinci, and of Dürer, must have played a great part in pointing the way towards a better era of herbal illustration, for after they had shown what *could* be done, it was scarcely possible for botanists to remain satisfied with the rudely conventional type of drawing found in the incunabula. Dürer has a special claim to be remembered by ecologists, for in each of his two coloured drawings of sods

203

VII. Botanical Illustration

of turf, a tangled group of growing plants, exactly as it occurred in nature, is portrayed with an unrivalled combination of artistic charm and scientific accuracy. In herbals *sensu stricto*, plants are almost always represented singly, so that their relatedness to one another, and to the habitat, is lost; but an interest in this relatedness occasionally appears. Fig. 95, p. 203, for instance, which is from a zoological book of

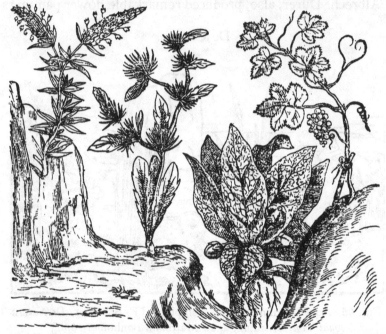

Fig. 96. Plants of varying altitudes [Porta, *Phytognomonica*, 1588]

1558, is definitely ecological in intention, since it represents an association of aquatic animals and marsh plants. An exceptional illustration from Porta's *Phytognomonica*, of 1588, may be included here also; it shows several plants growing in their appropriate habitats, which are condensed in a symbolic manner (fig. 96).

Though Dürer's drawings must have had a profound influence upon the pictures in botanical books, he did not con-

Plate xxi

Study of *Aquilegia vulgaris* L., Columbine [Albrecht Dürer, 1526.
Drawing in the Albertina, Vienna] *Reduced*

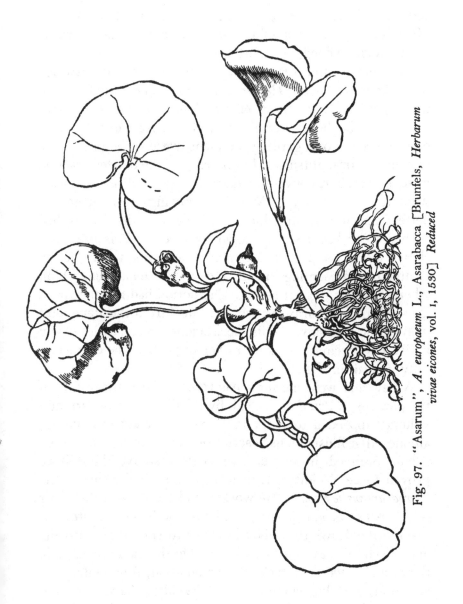

Fig. 97. "Asarum", *A. europaeum* L., Asarabacca [Brunfels, *Herbarum vivae eicones*, vol. i, 1530] *Reduced*

cern himself with these books directly. It was Hans Weiditz [Guiditius]—an artist of the same school, some of whose work was, indeed, formerly attributed to Dürer—who opened the botanical renaissance with the realistic drawings which he made for the *Herbarum vivae eicones* of Brunfels. The title—*Living portraits of plants*—indicates the distinctive feature of the book, namely that the artist went direct to nature, instead of regarding the plant world through the eyes of previous draughtsmen. In one respect the belated reaction from the earlier, conventionalised and generalised woodcuts went too far. Many of Weiditz' drawings were from imperfect specimens, in which, for instance, the leaves had withered, or had been damaged by insects. This is clearly shown in fig. 99, p. 208. The artist's ambition was evidently limited to representing the actual example he had before him, whether it was normal or not. The notion had not then been grasped, that the drawing which is ideal from the standpoint of systematic botany, avoids the accidental peculiarities of any individual specimen, seeking rather to portray the characters fully typical for the species.

A new and welcome light has recently been thrown upon the pictures in Brunfels' herbal, by the discovery at Bern of a volume, appended to Felix Platter's herbarium, containing, among other illustrations, a series of plant paintings in watercolour outlined in bistre. On critical study, these have proved to be Weiditz' original drawings for the cuts in the *Herbarum vivae eicones*. He worked on both sides of the paper —a luckless economy which led the methodical Platter to mangle the drawings cruelly in order to stick them into the appropriate places in his collection. On the backs of some of them, which were unstuck for examination, fragmentary inscriptions, in the handwriting of Weiditz, have been detected, including, in one place, the date "1529". A number of the drawings, which are of excellent quality, have now been published in facsimile. It seems evident that the intention of the colouring was not only to assist the craftsmen to translate the plant pictures in terms of black and white, but also to serve

Fig. 98. "Kuchenschell", *Anemone Pulsatilla* L., Pasque-flower [Brunfels, *Herbarum vivae eicones*, vol. I, 1530] *Reduced*

as a model for coloured copies issued by the publishers[1]. From the figures here reproduced, some idea can be obtained

[1] For this suggestion the writer is indebted to Dr T. A. Sprague.

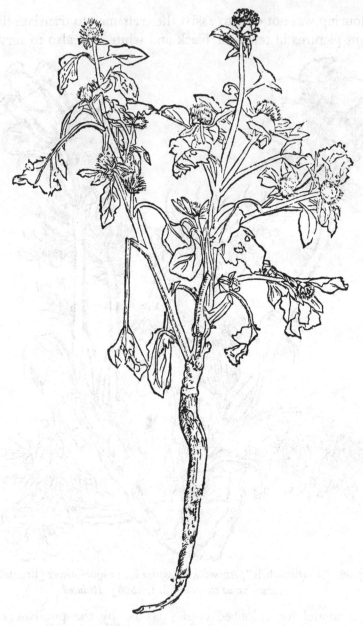

Fig. 99. "Lappa", *Arctium Lappa* L., Burdock [Brunfels, *Herbarum vivae eicones*, vol. II, 1531] *Reduced*

of the technique of the engraver who interpreted Weiditz' paintings. His outline is crisp, free, and virile, and little shading is employed.

Regarded from the point of view of decorative book illustration, the beautiful drawings of the botanical renaissance sometimes fail to reach the standard set by earlier work. The full black, velvety line of many of the fifteenth-century wood-engravings, and the occasional use of white in relief upon a solid black background, gives a sense of depth, which is specially compatible with black letter type. A page bearing such illustrations is often more satisfying to the eye than one in which the desire accurately to express the subtleties of plant form has led to the use of a more delicate line. Since, however, the primary object of the herbal artists was a scientific and not a decorative one, the gain in realism of the later work more than compensates for some loss in the harmonious balance of black and white.

Our chronological survey of botanical woodcuts brings us next to those published by Egenolph in 1533 to illustrate Rösslin's *Kreutterbüch*. The lesser-celandine (fig. 35, p. 72), the hart's-tongue-fern (fig. 100, p. 210), and the pictures on the title-page (fig. 34, p. 71), are reproduced here as examples. These woodcuts are of no real importance, since they are, for the most part, merely copies of those in the *Herbarum vivae eicones*, on a smaller scale and reversed; it is true that their general look is different, but this is because, as an expert has pointed out, Egenolph's engraver used a flowing, easy, almost brush-like line, very distinct from that of the craftsman who worked for Brunfels. To protect himself against Egenolph's piracy, the publisher, Schott, brought an action at law, and the culprit offered various shuffling excuses. There is no record of the actual result of the suit, but, in 1534, Schott published the *Kreüterbüch contrafayt*, a German edition of the *Herbarum vivae eicones*, illustrated, in the main, with the very cuts that had been used in Rösslin's

VII. Botanical Illustration

Kreutterbûch. This seems to indicate that Egenolph had been legally compelled to hand them over, and hence that this early action for infringement of copyright came to a successful issue.

It will be realised that, as the third part of Brunfels' work had not appeared when Egenolph's book was published,

Hirtʒʒung.
Scolopendria.   Linguacerui.

Fig. 100. "Scolopendria", Hart's-tongue-fern
[Rösslin (Rhodion), *Kreutterbûch*, 1533]

the latter must have been at a loss for figures of the plants which were not included in the first two volumes. We find that in the case of one such plant, the asparagus, he solved the problem by going back to the familiar old woodcut which had done duty in the *Herbarius zu Teutsch*, and in the *Ortus sanitatis* (fig. 71, p. 166).

After being deprived of Brunfels' pictures, Egenolph did not mend his ways; he merely turned his attention to other victims. In his later volumes he used woodcuts pirated from the

210

DIPSACVS ALBVS.   Weiß Gartendistel.

Fig. 101. "Dipsacus albus", *D. fullonum* L., Fullers'-teasel
[Fuchs, *De historia stirpium*, 1542] *Reduced*

books of Fuchs and Bock, whose engravings we must now consider.

In the work of Leonhart Fuchs plant illustration may perhaps be held to have reached an even higher level than in that of Brunfels. This was partly due to the perfect collaboration between the author and his draughtsmen, a collaboration which is known, from Brunfels' own statement, to have been lacking when the *Herbarum vivae eicones* was being prepared. The only failure of the kind which can be attributed to Fuchs and his staff is that the wood-cutter was not always entirely clear about the artist's intention. This, perhaps, is hypercriticism, for the result is, in general, admirable. At a later date, when the botanical significance of the detailed structure of the flower and fruit was recognised, figures were produced which conveyed more minute and complete information on these points than Fuchs' pictures; but, at least in the opinion of the present writer, the illustrations to Fuchs' herbals (*De historia stirpium*, 1542, and *New Kreüterbüch*, 1543) represent the high-water mark of that type of botanical drawing which seeks to express the individual character and habit of each species, treating the plant broadly as a whole, and not laying more stress upon the reproductive than the vegetative organs.

Fuchs' figures were on so large a scale that the plants frequently had to be curved in order to fit them even into his folio pages. The examples here reproduced (figs. 31, p. 66; 32, p. 68; 33, p. 69; 65, p. 154; 78, p. 175; 79, p. 178; 80, p. 180; 101, p. 211; 102, p. 213; 103, p. 214; 104, p. 216) do not convey an entirely just idea of the character of the work, since the narrow line employed in the woodcuts is thickened, relatively, by reduction. In the originals, however, the thinness of the line is not altogether an advantage, since it has a tendency to give a poor and slightly unfinished look to the less detailed of the engravings. It may be that Fuchs was allowing for the fact that the pictures in some copies

Fig. 102. "Apios", *Lathyrus tuberosus* L., Earth-nut-pea [Fuchs,
De historia stirpium, 1542] *Reduced*

213

PERSICA   Pferſichbaum.

Fig. 103. "Persica", *Prunus persica* Stokes, Peach [Fuchs, *De historia stirpium*, 1542] *Reduced*

would be coloured. Many of the extant specimens of fifteenth- and sixteenth-century herbals have the figures painted. It is sometimes difficult to get evidence as to the exact date of this colouring, but it was doubtless often done in the publisher's office; Christophe Plantin of Antwerp employed certain women illuminators to colour by hand the botanical books which he produced.

Sometimes in Fuchs' figures a peculiarly decorative spirit is shown, as in the earth-nut-pea (fig. 102, p. 213), which fills the rectangular space almost in the manner of an "all-over" wall-paper pattern. It must not be forgotten, when discussing woodcuts, that the artist who drew upon the block for the engraver was working under special conditions. It was impossible for him to be unmindful of the boundaries of the block, when these took the form, as it were, of miniature precipices under his hand. The writer has been told by her father—who, in the nineteenth century occasionally drew upon the wood for the engraver—that to avoid a rectangular appearance required a distinct effort of will. It is not surprising, under these circumstances, that the herbal illustrator who drew upon the block should often have been so much obsessed by its rectangularity that he accommodated his drawing to its form in a way that was unnecessary and far from realistic, though sometimes very attractive. This is exemplified in the earth-nut-pea, to which we have just referred, and also in figs. 44, p. 94; 47, p. 98; 51, p. 106; 55, p. 115; 111, p. 225. The traditional mode of representing trees, with an almost square crown, is probably a direct result of the form of the block. This can be seen particularly well in figs. 94, p. 200, and 108, p. 222; it is also indicated in fig. 17, p. 36—a peach-tree as viewed by the wood-cutter of the *Ortus sanitatis* in the fifteenth century—while traces of the same mannerism can be observed again in Fuchs' peach-tree (fig. 103, p. 214), despite its more naturalistic treatment. At the present day, when photographic methods of reproduction

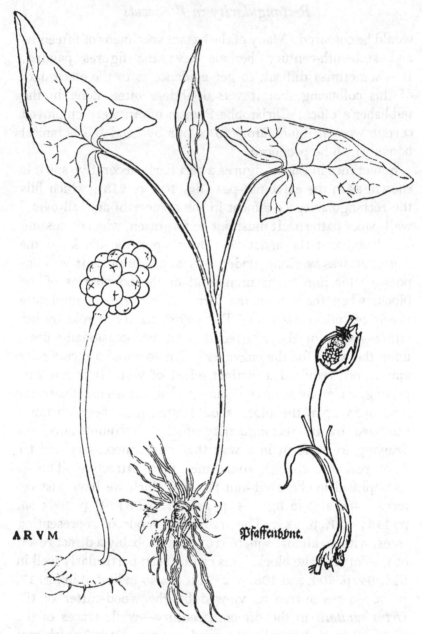

A R V M   Pfaffenbÿnt.

Fig. 104. "Arum", *A. maculatum* L., Lords-and-ladies [Fuchs, *De historia stirpium*, 1542] *Reduced*

are almost always employed, the artist is no longer so oppressively conscious of the boundaries of the space which his drawing will occupy.

The figures here reproduced show how great a variety of subjects received successful treatment in Fuchs' work. His elegant drawing of asparagus (fig. 78, p. 175) forms an interesting contrast with the symbolic version of the same plant in the *Ortus sanitatis* (fig. 71, p. 166), while the solid cabbage (fig. 31, p. 66) is realised in a way that reveals the intrinsic charm of this homely object. In lords-and-ladies (fig. 104, p. 216) the fruit, and a dissection of the inflorescence, are represented, so that, botanically, the drawing reaches a high level. Fuchs' woodcuts are nearly all original, but that of the white-water-lily appears to have been founded upon Brunfels' figure.

We have hitherto spoken, for the sake of brevity, as if Fuchs in fact executed the figures himself; but this was not so. His staff included Albrecht Meyer, who drew and perhaps coloured the plants from nature; Heinrich Füllmaurer, who copied the drawings on to the wood; and also an engraver, Veit Rudolf Speckle, who did the actual cutting of the blocks. Fuchs delighted to honour his colleagues, for, at the end of the book, there are portraits of all three (fig. 105, p. 218); the artist is drawing a plant, with a brush fixed in a quill.

The delineation and painting of flowers is sometimes dismissed almost contemptuously, as if it were a mere accomplishment, rather than an exacting art. The Chinese in early days knew better, for they held that a lifetime spent in making studies of bamboo alone, was not wasted. In Europe, also, the artists who have produced the happiest interpretations of flowers have often been men who also did admirable figure work. The examples of Leonardo da Vinci and of Albrecht Dürer need no stressing; and, within the herbals themselves, we have the portraits in *De historia stirpium* (fig. 30, p. 65; 105, p. 218), in Bock's *Kreuter Buch*

SCVLPTOR

Fig. 105. The Draughtsmen and the Engraver employed by Leonhart
Fuchs [*De historia stirpium*, 1542] *Reduced*

218

Fuchs' Wood-blocks

(fig. 26, p. 58), and in Columna's *Ekphrasis* (pl. x, f.p. 98), to remind us that the herbal artists could, when they were so disposed, achieve success in a more ambitious field.

That Fuchs realised his debt to the draughtsmen and engraver who worked for him, and also that he considered it his duty to supervise their labours in an unflinching spirit, is shown in the preface to *De historia stirpium*; his remarks upon the illustrations may be translated as follows:

"As far as concerns the pictures themselves, each of which is positively delineated according to the features and likeness of the living plants, we have taken peculiar care that they should be most perfect, and, moreover, we have devoted the greatest diligence to secure that every plant should be depicted with its own roots, stalks, leaves, flowers, seeds and fruits. Furthermore we have purposely and deliberately avoided the obliteration of the natural form of the plants by shadows, and other less necessary things, by which the delineators sometimes try to win artistic glory: and we have not allowed the craftsmen so to indulge their whims as to cause the drawing not to correspond accurately to the truth. Veit Rudolf Speckle, by far the best engraver of Strasburg, has admirably copied the wonderful industry of the draughtsmen, and has with such excellent craft expressed in his engraving the features of each drawing, that he seems to have contended with the draughtsman for glory and victory."

The actual wood-blocks from which the figures in *De historia stirpium* were printed, survived for a very long time. They were used in Zurich by Salomon Schinz to illustrate a book called *Anleitung zu der Pflanzenkenntniss*, which appeared in 1774, no less than 232 years after the first publication of the cuts. We have already mentioned the extensive copying, in later herbals, of the smaller woodcuts in Fuchs' octavo edition (p. 70). Jerome Bock made some use of them, as well as of the pictures in the *Herbarum vivae eicones*, in the second edition of his *Kreuter Büch* (1546), which was

219

Von Winter grün.

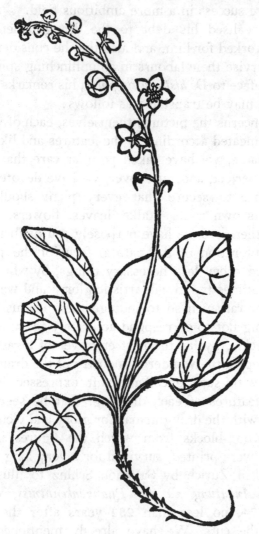

Fig. 106. "Winter grün", *Pyrola*, Winter-green
[Bock, *Kreuter Bůch*, 1546]

the next important illustrated botanical work to appear after Fuchs' herbal. There was, nevertheless, a strong element of originality in Bock's pictures. The artist whom he employed —David Kandel, a young lad, son of a burgher of Strasburg— produced engravings which are often of interest, even apart from their botanical aspect. For instance, the woodcut of an oak-tree includes a swineherd with his swine; the chestnut gives occasion for a hedgehog (fig. 108, p. 222); and, in another block, a monkey and several rabbits are introduced, one of the latter holding a shield bearing the artist's initials. These recur in the portrait of Bock (fig. 26, p. 58), which reveals Kandel's gift both for portraiture and decoration. The woodcut of *Trapa*, the bull-nut (fig. 29, p. 63) is a highly imaginative production, which shows clearly that neither artist nor author can ever have seen the plant. This absurd picture had a long life. It appeared at as late a date as 1783, in the last edition of Lonitzer's herbal.

The illustrations in Mattioli's *Commentarii* differ markedly in

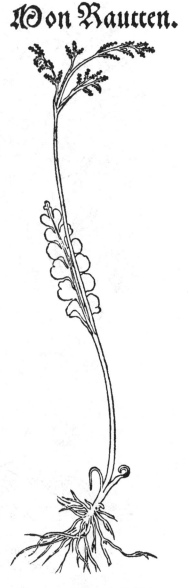

Von Rautten.

Fig. 107. "*Lunaria*", *Botrychium*, Moonwort [Bock, *Kreuter Büch*, 1546]

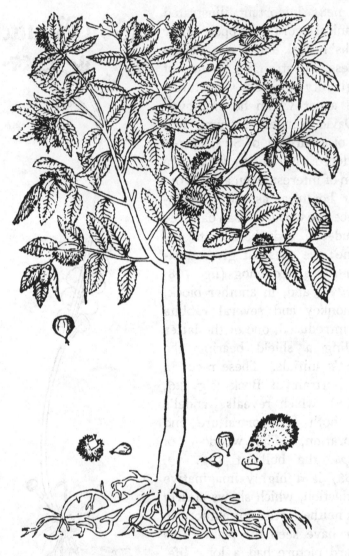

Fig. 108. "Castanien nuss", *Castanea*, Chestnut
[Bock, *Kreuter Bůch*, 1546]

style from those in the herbals which we have hitherto named. Details, such as veins and hairs, are often worked out elaborately, while shading is much used, a considerable mastery of parallel lines being shown. In its earlier form the book had small figures (e.g. figs. 44, p. 94; 45, p. 95; 109, p. 223; 110, p. 224); the cut representing fungi (fig. 109) gains a certain fascination from the low eye level, which suggests that one is seeing the toadstools from the standpoint of the

Fig. 109. "Fungi", Toadstools [Mattioli, *Commentarii*, 1560] *Reduced*

snail in the foreground. These pictures, like those of Fuchs, had a prolonged existence; they were even copied in a little French herbal of 1766, more than two hundred years after their first appearance. A later version of Mattioli's *Commentarii*, with different and larger illustrations, appeared first in Prague in 1562, and later in Venice. Examples from the Venetian edition of 1565 are shown on a reduced scale in figs. 46, p. 96; 47, p. 98; 111, p. 225. These woodcuts resemble the smaller ones in character, but are more decorative in effect, and often remarkably fine. Whereas in the work of Brunfels and Fuchs, the curve of a single stalk may afford the key-note to the whole drawing, in the work of Mattioli, the eye most frequently finds its satisfaction in the

rich massing of foliage, fruit, and flowers, suggestive of
southern luxuriance. Many of his figures would require little

ROSACEVM

Fig. 110. "Rosaceum" [Mattioli, *Commentarii*, 1560]
Reduced

modification to form the basis of a tapestry pattern. In
fig. 112, p. 226, the picture of the mulberry from the Prague
edition is reproduced. It includes the wood-cutter's initials,

Mattioli's Woodcuts

with his device of a graving tool, which is shown on a larger
scale in fig. 113, p. 226. A bibliographer, who has made a

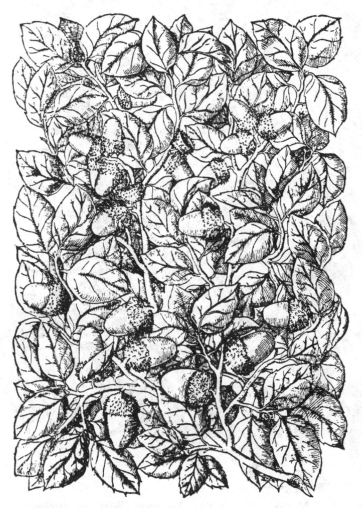

Fig. 111. "Suber Primus" [Mattioli, *Commentarii*, 1565]
Reduced

special study of this rare Bohemian herbal, notes that, in a
treatise on monograms, this burin device has been learnedly
misdescribed as "un crocodile".

AH   225   15

Fig. 112. "Morus", *M. nigra* L., Mulberry [Mattioli, *Herbarz: ginak Bylinář*, 1562] *Reduced*

Fig. 113. Initials and device of graver's burin [from fig. 112] *Enlarged*

226

We have now to consider the wood-engravings which illustrate the work of the three Low Country herbalists—

Fig. 114. "Tragorchis", *Orchis hircina* L., Lizard-orchis
[Dodoens, *Pemptades*, 1583]

Dodoens, de l'Écluse, and de l'Obel. In the original edition of Dodoens' Flemish herbal (*Crüÿdeboeck*, produced by Van der Loe in 1554), more than three-quarters of the pictures, as the author himself explains, were derived from Fuchs'

15-2

octavo edition of 1545, while the remainder, of which fig. 58, p. 126, is an example, were original. Further blocks were

Aconitum luteum minus.

Fig. 115. "Aconitum luteum minus", *Eranthis hiemalis* L., Winter-aconite [Dodoens, *Pemptades*, 1583]

added in later editions. All the botanical books which Dodoens wrote after this Flemish herbal were produced by Christophe Plantin. Plantin was also the publisher for de l'Écluse and de l'Obel, and as their blocks were held more or less as a common stock, it is best to treat them, for historical purposes, as forming a single group.

The first book by Dodoens for which Plantin was responsible, was the *Frumentorum historia* of 1566. From the old records of the *Officina Plantiniana*, we learn that nearly all the illustrations in this book were drawn by Pierre Van der Borcht, and that the artist was paid at the rate of 12 or 13 sous for a drawing. For another of Dodoens' books, *Florum, et coronarium herbarum historia* (1568), Arnaud Nicolai, the engraver who cut the blocks, received 7 sous apiece.

After Dodoens had published several small works with Plantin, he became desirous of collecting his writings into a comprehensive Latin herbal. Towards the illustration of this volume (*Stirpium historiae pemptades sex*, 1583) Plantin was able, in 1581, after the death of Jean Van der Loe, to buy from his widow all the blocks which had been used in the Flemish herbal; to these were added a number of new ones, made under the direction of Dodoens. Examples of the figures in the *Pemptades* are shown in figs. 40, p. 83; 41, p. 84; 114, p. 227; 115, p. 228.

No special discussion of de l'Obel's figures is necessary, since they partook of the character of the rest of the woodcuts for which Plantin made himself responsible. Mistletoe (fig. 64, p. 151), yellow-water-lily (fig. 76, p. 172), and tobacco (fig. 116, p. 230), are given here as examples.

The woodcuts illustrating the works of de l'Écluse are perhaps the best of those associated with this trio of botanists. De l'Écluse often took pains to include fruiting as well as flowering stages, thus increasing the botanical value of his figures, three of which are shown in figs. 42, p. 87; 66, p. 156; 117, p. 231. One of the treasures of the Preussische Staatsbibliothek of Berlin is a collection of 1856 plant drawings in water-colour, which have been discovered from internal evidence to have been executed under the direction of de l'Écluse, many of them by Pierre Van der Borcht. Amongst other pictures, this collection includes the originals of the woodcuts in de l'Écluse's book on the plants of Spain. A

Fig. 116. "Tabaco Nicotiane", *Nicotiana Tabacum* L., Tobacco [de l'Obel, *Kruydtboeck*, 1581. This figure appeared originally in Pena and de l'Obel, *Stirpium adversaria nova*, 1570,1]

reduced reproduction of the water-colour of the dragon-tree, which de l'Écluse saw in the garden of the monastery of "S. Maria a Gratia" at Lisbon, is shown in pl. xxii for comparison with the woodcut (fig. 117, p. 231), which was made from it. This is the first known figure of *Dracaena* in a

Plate xxii

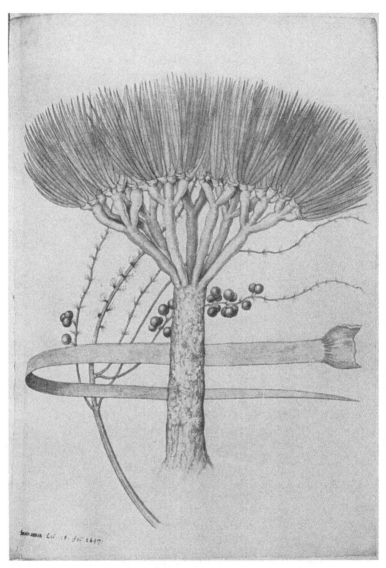

Dracaena Draco L., Dragon-tree

[From a water-colour in the Clusius collection, Preussische Staatsbibliothek, Berlin, Handschriften-Abteilung, Libri picturati, A 16–33]
Reduced

botanical book, but about a hundred years earlier it had been
represented in a copperplate of *The Flight into Egypt*, by

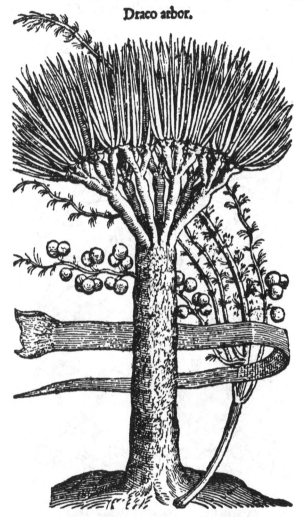

Fig. 117. "Draco arbor", *Dracaena Draco* L., Dragon-tree
[de l'Écluse, *Rariorum...per Hispanias*, 1576]

Martin Schongauer. This engraving, which is believed to
belong to the period 1469–74, includes a palm-tree, bent down
by angels, so that St Joseph can gather the dates, and, beside

231

it, a dragon-tree, with two engaging little lizards, one ascending and one descending the trunk.

The popularity of the large collection of blocks assembled

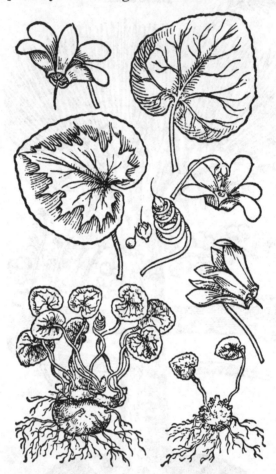

Fig. 118. "Cyclaminus", Cyclamen [Camerarius (Mattioli), *De plantis Epitome*, 1586]

by the publishing house of Plantin is shown by the frequency with which they were copied. It has been pointed out that the woodcut of a clematis, which was first seen in Dodoens' *Pemptades* of 1583, reappears either in identical form, or more or less accurately copied in works by de l'Obel, de l'Écluse, Gerard, Parkinson, Jean Bauhin, Chabraeus, and Petiver. Plantin's actual collection of blocks seems to have been used for the last time when Johnson's edition of Gerard's herbal made its final appearance in 1636.

Fig. 119. "Rosa Hierichuntica", *Anastatica hierochuntica* L., Rose-of-Jericho [Camerarius, *Hortus medicus*, 1588]

Another school of plant illustration is represented by Gesner and Camerarius. We have traced the history of Gesner's drawings on pp. 111–113, and shown the impossibility of separating the work of these two botanists. An excellent feature of their pictures is that they often include detailed analyses of vegetative and floral characters, and studies of

233

VII. Botanical Illustration

different phases in the life history (e.g. fig. 118, p. 232; and also figs. 37, p. 77; 54, p. 112; 81, p. 182; 82, p. 183). In fig. 119, p. 233, the seedling of the rose-of-Jericho is drawn side by side with the mature plant, while in fig. 38, p. 78, the appearance of a sprouting date stone is shown with great clearness. This attention to seedlings foreshadows a branch of study which has come to its own in modern botany. The idea that the perfect illustration should include representations of the plant at every stage of its life history is no new one. Pliny, as long ago as the first century of our era, complained (in the words of Philemon Holland's translation) that the herbalists' pictures "came far short of the marke, setting out hearbs as they did at one onely season (to wit, either in their floure, or in seed time) for they chaunge and alter their forme and shape everie quarter of the yeere".

A number of wood-blocks were used at Lyons to illustrate d'Aléchamps' massive work, the *Historia generalis plantarum*, 1586,7. Many of these figures were taken from the herbals of Mattioli, Fuchs, and Dodoens. The blocks cut at Lyons were apt to be embellished with representations of insects and detached flowers, scattered over the background with no apparent object except to fill the space (e.g. fig. 57, p. 118). These decorations are found in a French version of Mattioli's *Commentarii* made by Desmoulins in 1572, which provided some of the pictures used in d'Aléchamps' herbal.

Among other botanical wood-engravings of minor importance, we may mention those in the works of Pierre Belon, a contemporary and friend of Valerius Cordus. In his *De arboribus* (1553) there are some pleasant woodcuts of trees, one of which is reproduced in fig. 120, p. 235. The initial letters used in the present volume, are taken from another of Belon's books.[1] Specimens of the amusing little illustrations to Castor Durante's *Herbario nuovo*, of 1585,

---

[1] Pierre Belon, *Les Observations de plusieurs singularitez et choses memorables....* Paris, 1553.

234

are shown in figs. 50, p. 103; 77, p. 174; 121, p. 236; some of them are derived from Mattioli.

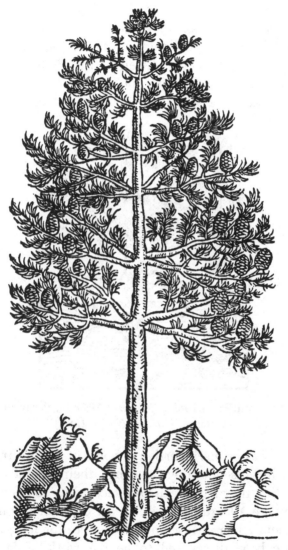

Fig. 120. "Cedrus", Cedar [Belon, *De arboribus*, 1553]

There is a certain small book—a great rarity—containing some charming woodcuts of plants, which gains interest for

235

us since it was produced in England. This is *La Clef des Champs* (1586), by Jacques Le Moyne de Morgues, who accompanied one of the French expeditions to America, and, after his return, settled in Blackfriars, where he occupied himself with painting and engraving. At one time he was in the service of Sir Walter Raleigh; Sir Philip Sidney was numbered among his friends. The purpose of the book, which he undertook, as he tells us, "volontiers et selon mon petit

Fig. 121. "Lentisco del Peru", *Schinus Molle* L., Peruvian-mastich
[Durante, *Herbario Nuovo*, 1585]

pouvoir", was primarily artistic. Some of the illustrations in the British Museum copy have had their outlines pricked, showing that they have been used, according to his intention, for "ouvrage à l'éguille". A fine set of water-colours in the Victoria and Albert Museum, which were for long unidentified, have been discovered to be the work of Jacques Le Moyne, and to include some of the originals from which the woodcuts in his book were derived.

The engravings in Porta's *Phytognomonica* (1588), and in

236

Prospero Alpino's book on Egyptian plants (1592), are of good quality. One of Porta's figures (fig. 96, p. 204) has already been mentioned; some other curious examples, the significance of which will be considered in the next chapter, are seen in figs. 127, p. 253, and 128, p. 257. The glasswort, which is among the best of the woodcuts in Alpino's book, is reproduced in fig. 49, p. 101.

Passing on to the seventeenth century, we find that the *Prodromos* of Gaspard Bauhin (1620) contains a number of original illustrations, but they are not very remarkable, and often have rather the appearance of having been drawn from pressed specimens. Two examples of these woodcuts will be found in figs. 55, p. 115, and 68, p. 159.

Parkinson's *Paradisus Terrestris* of 1629 includes a considerable proportion of new figures, besides others borrowed from previous writers. The engravings were made in this country by Switzer. Artistically they are poor, and the innovation of including a number of species in one large woodcut is not very successful. Fig. 61, p. 137, shows a twig of barberry, which is but a single item in one of these comprehensive illustrations.

Among still later wood-engravings, we may mention the rather coarse cuts in the *Dendrologia*, published under Aldrovandi's name more than sixty years after his death; one of these pictures, that of the orange, or "Mala Aurantia Chinensia", is reproduced in fig. 122, p. 238, on a greatly reduced scale.

In the present chapter no attempt has been made to discuss the illustrations of those herbals (e.g. the works of Turner, Tabernaemontanus, Gerard, etc.) in which most of the woodcuts are copied from previous books. In the majority of such cases, the source of the figures has already been indicated in Chapter IV. Something must be said, however, about a few of the pictures in Johnson's editions of Gerard's *Herball* (1633 and 1636), which are interesting, not on account of any novelty, but from their extreme antiquity. One of these is a conventional cut, which is described as the "Saxifrage of the

[Fig. 122. "Mala Aurantia Chinensia", Orange
[Aldrovandi, *Dendrologia*, 1667,8] *Reduced*

ancients", but which is, in reality, wholly unidentifiable. Johnson explains that his friend, John Goodyer, sent him the drawing, which he had copied out of a manuscript of Apuleius. Other pictures in the herbal have a known ancestry which goes back still further. Dodoens, in his *Pemptades*, had included a certain number of figures which he labelled "Ex Codice Caesareo", since they were taken from the Anicia Juliana codex of Dioscorides (A.D. 512). Johnson, who used the collection of blocks assembled by Plantin to illustrate the work of the Low Country herbalists, acquired, among the rest, these copies of Greek originals. When he printed them, they may have been more than sixteen hundred years old, since they probably dated back to the days of Krateuas, in the first century before Christ. The woodcut called "Coronopus" has been chosen from this ancient group for reproduction here (fig. 123, p. 240), side by side with its source in the Anicia Juliana codex (pl. xxiii, f.p. 240). Johnson singles it out for special approval. He observes that Dodoens had it "out of an old Manuscript in the Emperors Library", and, after some slight criticism of it, adds: "yet all the parts are well exprest, according to the drawing of those times, for you shall finde few antient expressions come so neere as this dothe". Despite his praise, "Coronopus" is scarcely identifiable. The Anicia Juliana drawings are probably copies of still earlier work, and this picture is perhaps degraded from an ancestral figure of *Lotus ornithopodioides* L.

Before we pass on to the next phase in herbal illustration, it may be well to look back over the period of wood-engravings. Some impression of the evolution of the botanical cut during the first hundred years in the life of the printed herbal [1481–1581] may be obtained by comparing the delineations of the same type of plant in a number of books. Pictures of the water-lilies have been chosen here for this purpose. We begin with the crude and absurd drawing of "Nymfea" in a ninth-century manuscript of the *Herbarium* of Apuleius Platonicus (fig. 72, p. 167), beside which is placed the corresponding cut from the printed book (fig. 73). Fig. 21,

± 6 *Coronopus ex codice Cæsareo.*
Crow-foot Trefoile.

Fig. 123. "Coronopus" [Johnson's edition of
Gerard's *Herball*, 1633]

240

Plate xxiii

ΚΟΡѠΝΟΠΟΥC, possibly *Lotus ornithopodioides* L. [Dioscorides, *Codex Aniciae Julianae* (Vind. Med. Gr. I), circa A.D. 512, facsimile, 178 verso] *Reduced*

p. 49, though it is a sixteenth-century woodcut, is based upon that of a water-lily in the *Herbarius zu Teutsch* (1485), while fig. 74, p. 169, is another conventional and symbolic picture of the same plant, with the mannerisms of the incunabula. In fig. 75, p. 170, we have said farewell to fifteenth-century methods, and we pass by a sudden transition to Brunfels' graceful and naturalistic engraving of the white-water-lily, which belongs to a wholly different category from its predecessors. In the "Ninfea" of Durante (fig. 77, p. 174), the plants are seen in relation to their surroundings; this picture is earlier than its date would lead one to suppose, since it is copied on a reduced scale from Mattioli's illustration of *Nymphaea* and *Nuphar*. The yellow-water-lily is shown again in fig. 76, p. 172, which is one of de l'Obel's cuts; it is a good example of the compact decision and solidity of Plantin's work, but it lacks the delicate air of Brunfels' picture.

If we disregard the incunabula, and consider only the period in the sixteenth century when the printed herbal was at its best from the standpoint of illustration, we find that the number of sets of engravings which were pre-eminent— either on account of their intrinsic qualities, or because they were repeatedly copied from book to book—was strictly limited. We might almost say that there were only five collections of wood-blocks which were of first-rate importance—those namely of Brunfels, Fuchs, Mattioli, Plantin, and Gesner-Camerarius—all of which were produced in the sixty years between 1530 and 1590.

At the close of the sixteenth century, the art of wood-cutting was distinctly on the wane, and had begun to be superseded by engraving on metal. There was some botanical advantage in this change, since minute detail can be expressed more realistically in incised metal. Moreover, in the early days of botanical etching and engraving, one man, as a rule, both made the drawing and worked upon the plate;

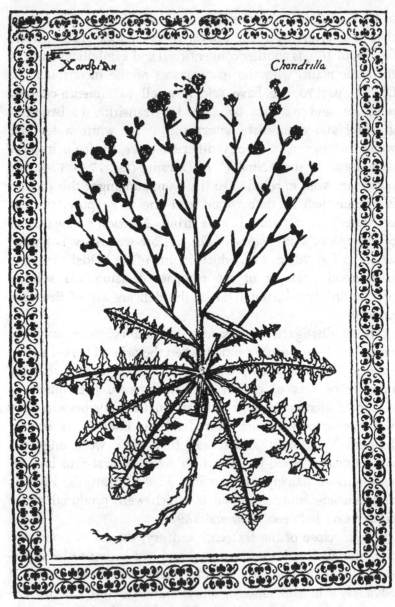

Fig. 124. "Chondrilla" [Columna, *Phytobasanos*, 1592]

this prevented the misunderstandings sometimes noticed in wood-blocks, in which the final result might be due to as many as three different men, who did not always collaborate well. The earliest botanical book in which copper plate etchings were used, is believed to have been Fabius Columna's *Phytobasanos* of 1592. These etchings, two of which are shown in figs. 48, p. 99 and 124, p. 242, are accurate and pleasing. Both in the *Phytobasanos*, and in a later work, the *Ekphrasis*, the structure of the flower and fruit is often shown separately; the figures, in this respect, are comparable with those of Gesner and Camerarius, though, owing to the smallness of their scale, they do not convey so much information. In the nineteenth century, the original drawings for the *Ekphrasis* came to light in a Neapolitan collection. It is believed that Columna not only drew the plants himself, but also etched them. He expressly mentions that he used wild plants as models wherever possible, because cultivation is apt to produce alterations in the form. The decorative border surrounding each of the figures is not part of the etching, but is printed with the type, an inconsistency which is aesthetically successful, in defiance of the laws of taste. If we accept the theory that the etching reproduced in pl. x, f.p. 98, is a self-portrait, we must conclude that Columna was a considerable artist.

In the seventeenth century, a large number of botanical books were illustrated by means of copperplates. The majority of these are not essentially herbals, so they scarcely come within our purview. We will limit ourselves to a glance at a group belonging to the thirty years from 1611 to 1641, which seems to have been a particularly fine period in plant delineation. In 1611 Paul Reneaulme's *Specimen Historiae Plantarum* was published in Paris; though it was illustrated with good engravings, the effect was somewhat marred by the transparency of the paper. The succeeding year saw a book containing very attractive pictures of garden plants, the *Florilegium Novum* of Johann Theodor de Bry; the same engraver was responsible for the portrait of Gaspard Bauhin, reproduced in pl. xiii, f.p. 114. In 1613 appeared Besler's *Hortus Eystettensis*, called by Sir Thomas Browne "the massiest" of the herbals; its spacious engravings measure about 18½ by 15¾ inches within the platemarks. Several species are sometimes included in one plate, while, in others, advantage is taken of the unusual area to do justice to the luxuriant growth of a single specimen. Many of the pictures are admirable. In the following year, 1614, a book was published which contained particularly delicate copperplates of plants, and which is now available in a twentieth-century version. This is the *Hortus floridus* of Crispian de Passe [Crispijn vande Pas] the Younger, a member of a renowned family of engravers. In 1615, an English edition of this work was published at Utrecht, under the title of *A Garden of Flowers*. The plates are the same as those in the first part of the original work. The engraver was particularly successful with the bulbous and tuberous plants which have long been a specialty of Holland. The soil in which the plants grow is often shown, the artist's eye-level being so low that the flowers stand up against the sky. This mannerism is characteristic, not only of the plant drawings of the Dutch, but also of their early landscapes; it seems to be a natural reaction to life in a flat country. Pl. xxiv is a typical example of de Passe's work, but only part of the original illustration is here reproduced.

The purchaser of *A Garden of Flowers* receives detailed directions for the painting of the figures, which he is expected to accomplish himself. The book is divided into four parts, appropriate to the four seasons, and each part is preceded by an encouraging verse, intended to keep alive the owner's enthusiasm for his task. The stanza at the beginning of the last section reveals some anxiety on the part of the author, lest the amateur artist should have begun to weary over the

Plate xxiv

"Crocus Byzantinus" and "Crocus Montanus hispan." [Part of a plate from Crispian de Passe, *Hortus floridus,* 1614]

lengthy occupation of colouring the plates. It reads as follows:

> If hethertoe (my frende) you have,
> Performde the taske in hand:
> With ioy proceede, this last will be
> The best, when all is scande.

In the sixteen-twenties, the *Theatrum Florae* was published anonymously in Paris.[1] Its title-page, with allegorical figures, is rendered particularly striking by a rich dark background. Among the more interesting of its plates is a delicately elaborate drawing of the feathered-hyacinth. An album of flower paintings by Daniel Rabel, dated 1624, has been discovered to contain the originals of the plates in the *Theatrum Florae*, so that their authorship is now established beyond doubt.

Among later copperplates, we will mention only the beautiful engravings in Matthew Merian's edition of de Bry's *Florilegium Novum*, which appeared in 1641.

In reviewing the illustrated botanical books of the first half of the seventeenth century, we realise that the volumes containing engraved plates tended to become picture-books, in which the text was reduced to a minimum. The fact is that, from its very nature, metal-engraving does not fall easily into place as a mode of book illustration. In wood-cutting, the lines are raised, and the method of printing is thus exactly the same as for type, while in metal-engraving, the lines are incised, so that the process is reversed. As a result of this difference, in a book with woodcuts there is no difficulty in keeping a just balance between text and illustrations, which are printed together: on the other hand, in a book with metal engravings, which have to be printed separately from the type in the form of relatively expensive plates, the pictures are liable to become the primary feature of the work, and to lose their relation to the text, which then

---

[1] The writer has seen only the edition of 1633.

245

declines. The consequence is that books with fine engravings appeal rather to the wealthy amateur, whose influence is patent in the seventeenth-century botanical books, in which the theme shows a definite shift towards horticulture. The humbler woodcut—or its modern representative in which photography is called into play—can be kept under firmer control, and is thus better fitted for the purposes of the working botanist. In the history of the science, illustration has, indeed, been the best of servants, but progress is inhibited when it usurps the master's place.

Chapter VIII

THE DOCTRINE OF SIGNATURES, AND ASTROLOGICAL BOTANY

uring the preceding chapters, we have restricted our discussion to those writings which may be credited with having taken some part, however slight, in advancing the knowledge of plants. We have, as it were, confined our attention to the main stream of botanical progress, and its tributaries; but before concluding, it may be well to call to mind the existence of more than one backwater, connected indeed with the main channel, but leading nowhere.

The subject of the superstitions with which herb collecting has been hedged about at different periods, belongs to the province of the student of folk-lore, and cannot be dealt with in detail in the present book. In earlier chapters we have touched upon the observances with which the Greek herb gatherers surrounded their calling, and the mysterious dangers, which are described in the *Herbarium* of Apuleius, as attending the uprooting of the mandrake. Though there is comparatively little reference to such matters in the works of the genuine herbalists of the sixteenth and seventeenth centuries, there is a succession of contemporary books of a different type, dealing with two pseudo-scientific themes— the "signatures" of plants, and botanical astrology. These works have a certain interest, rather on account of the surprising light which they throw upon the attitude of mind of their writers (and presumably of their readers also), than

247

from any intrinsic merit. One of these authors, in his preface, speaks of the "Notions" and "Observations" contained in his work, "Most of which I am confident are true, and if there be any that are not so, yet they are pleasant." The fact that the "Notions", cherished by the more credulous herbalists were "pleasant" (in the old sense of "amusing"), even if untrue, may perhaps justify a very brief discussion of their salient points.

Fig. 125. Title-page device of a Mandrake [Brunfels, *Contrafayt Kreüterbůch*, Ander Teyl, 1537]

The most famous of those heterodox writers who turned their attention to botany, was Theophrastus Bombast von Hohenheim (1493–1541), who latinised his name to Paracelsus; his portrait is shown in fig. 126, p. 249. He was a doctor, as his father had been before him, and after wide travel and much experience, he became professor at Basle. Here he gave serious offence by lecturing in the vulgar tongue, by burning the writings of Avicenna and Galen, and by interpreting his own works instead of those of the ancients. His disregard of cherished traditions, and his personal peculiarities, led to difficulties with his colleagues, and he only held his post for a very short time. For the rest of his life he was a wanderer on the face of the earth, and he died in comparative poverty at Salzburg in 1541.

The character and writings of Paracelsus are full of seemingly irreconcilable contradictions. Browning's poem

Fig. 126. Theophrastus von Hohenheim, called Paracelsus (1493–1541) [From a medal, see Weber, F. P., Appendix II]

perhaps gives a better idea of his career than any prose account aiming at factual accuracy. His life was so strange that the imagination of a poet is needed to revitalise it for us to-day. His almost incredible boastfulness is the main characteristic that everyone remembers; the word "bombast" was, indeed, formerly believed to have been coined from his name. In one of his works, after contemptuously dismissing all the great physicians who had preceded him—Galen, Avicenna and others—he remarks, "I shall be the Monarch and mine shall the monarchy be." The conclusion that there was an element of quackery in his composition can hardly be avoided, but it is also true that he was visited by flashes of scientific and mystical insight, and that he played a part in infusing new life into chemistry and medicine. Probably we shall find no fairer summary than that of Guy de la Brosse, who, in 1628, wrote of Paracelsus: "j'ai bien apperçeu qu'il a de tres-belles et tres-rares pensees, mais aussi qu'elles ne sont pas tousiours esgales."

Paracelsus' actual knowledge of botany appears to have been meagre, for not more than a couple of dozen plant names are found in his works. To understand his views on the properties of plants it is necessary to turn for a moment to his chemical theories. He regarded "sulphur", "salt", and "mercury" as the three fundamental principles of all bodies. The sense in which he uses these terms is symbolic, and thus differs entirely from that in which they are employed to-day. "Sulphur" appears to embody the ideas of change, combustibility, volatilisation and growth; "salt", those of stability and non-inflammability; "mercury", that of fluidity. The "virtues" of plants depend, according to Paracelsus, upon the proportions in which they contain these three principles.

The medicinal properties of plants are thus the outcome of qualities that are not obvious at sight. How, then, is the physician to be guided in selecting herbal remedies to cure the several ailments of his patients? In answer to this question Paracelsus adopted and expanded the ancient belief in the *Doctrine of Signatures.*

According to this doctrine, many medicinal herbs are stamped, as it were, with some clear indication of their uses. This may perhaps be best understood by means of a quotation from a seventeenth-century *Dispensatory,* described as a translation from Paracelsus. The powers of *Hypericum* are deduced as follows: "I have oft-times declared, how by the outward shapes and qualities of things we may know their inward Vertues, which God hath put in them for the good of

250

man. So in St. Johns wort, we may take notice of the form of the leaves and flowers, the porosity of the leaves, the Veins. 1. The porositie or holes in the leaves, signifie to us, that this herb helps both inward and outward holes or cuts in the skin.... 2. The flowers of Saint Johns wort, when they are putrified, they are like blood; which teacheth us, that this herb is good for wounds, to close them and fill them up."

A later devotee of the doctrine of signatures, who presented it with highly ingenious pseudo-scientific plausibility, was Giambattista Porta, who was born at Naples, probably a short time before the death of Paracelsus. He wrote a book about human physiognomy, in which he endeavoured to find, in the bodily form of man, indications as to his character and spiritual qualities. This study suggested to him the idea that the inner qualities, and the healing powers, of the herbs might also be revealed by external signs, and thus led to his remarkable work, the *Phytognomonica*, which was first published at Naples in 1588.

Porta developed his theory in detail, and pushed it to great lengths. He supposed, for example, that long-lived plants would lengthen a man's life, while short-lived plants would abbreviate it. He held that herbs with a yellow sap would cure jaundice, while those whose surface was rough to the touch would heal those diseases that destroy the normal smoothness of the skin. The resemblance of certain plants to certain animals opened to Porta a vast field for dogmatism on a basis of conjecture. Plants with flowers shaped like butterflies would, he supposed, cure the bites of insects, while those whose roots or fruits had a jointed appearance, and thus remotely suggested a scorpion, must necessarily be sovereign remedies for the sting of that creature.

The illustrations of the *Phytognomonica* are helpful in interpreting Porta's point of view. The part of man's body which is healed by a particular herb, or the animal whose bites or stings can be cured by it, are represented in the same

251

woodcut as the herb. For example, the back view of a human head with a thick crop of hair is introduced into the block with the maidenhair—a capacity for curing baldness being suggested by the hair-like delicacy of the leaf-stalks of this fern. A pomegranate with its seeds exposed, and a plant of "toothwort", with its hard white scale-leaves, are represented in the same figure as a set of human teeth; while a spotted snake, with double tongue, accompanies a drawing of aroids with their spotted stalks. A scorpion completes a picture of plants with articulated seed-vessels (fig. 127, p. 253); a shoot of heliotrope is also included, since, to Porta's vivid imagination its curved flower spike recalled a scorpion's tail.

It would serve little purpose to deal in detail with the various exponents of views similar to Porta's, such, for instance, as Johann Popp, who, in 1625, published a herbal written from this standpoint, and containing also some astrological botany. We will only now refer to one of the later champions of the doctrine of signatures, an English herbalist, who made that fantastic realm peculiarly his own. This was William Cole (whose name is wrongly spelt "Coles" on the title-pages of his works), a Fellow of New College, Oxford, who lived and botanised at Putney in Surrey. He seems to have been a person of much character, and his vigorous arguments would often be telling, were it possible to admit the soundness of his premisses.

William Cole carried the doctrine of signatures to as extreme a point as can well be imagined. His account of the Walnut, from his work *Adam in Eden* (1657), may be quoted as an illustration: "*Wall-nuts* have the perfect Signature of the Head: The outer husk or green Covering, represent the *Pericranium*, or outward skin of the skull, whereon the hair groweth, and therefore salt made of those husks or barks, are exceeding good for wounds in the head. The inner wooddy shell hath the Signature of the Skull, and the little yellow skin, or Peel, that covereth the Kernell of the hard *Meninga* and *Pia-mater*, which are the thin scarfes that envelope the brain. The *Kernel* hath the very figure of the Brain, and therefore it is very profitable for the Brain, and resists poysons;

Fig. 127. Herbs of the Scorpion [Porta, *Phytognomonica*, 1591]

For if the Kernel be bruised, and moystned with the quintessence of Wine, and laid upon the Crown of the Head, it comforts the brain and head mightily."

In Cole's writings we meet with instances of a particular ambiguity of thought which characterised the doctrine of signatures. The signature in some cases represents an animal injurious to man, and is taken to denote that the plant in question will cure its bites or stings; for instance, in *The Art of Simpling* (1656) we learn that "That Plant that is called *Adders tongue*, because the stalke of it represents one, is a soveraigne wound Herbe to cure the biting of an Adder." In other cases, the signature represents one of the organs of the human body, and indicates that the plant will cure diseases of that organ; for example, "*Heart Treyfoyle* is so called, not onely because the Leafe is Triangular like the Heart of a Man, but also because each Leafe containes the perfect Icon of an Heart, and that in its proper colour, *viz.* a flesh colour. It defendeth the Heart against the noisome vapour of the Spleen."

Cole did, at least, endeavour to follow his theories to their logical conclusion. He was much exercised because a large proportion of the plants with undoubted medicinal virtues have no obvious signatures. He concluded that a certain number were endowed with signatures, in order to set man on the right track in his search for herbal remedies, while the remainder were purposely left blank, to encourage his skill in discovering their properties for himself. An argument, plagiarised from Henry More, is that a number of plants are left without signatures, because if all were signed, "the rarity of it, which is the delight, would be taken away by too much harping upon one string".

Our author was evidently a keen and enthusiastic collector of herbs. He complains bitterly that physicians leave the gathering of herbs to the apothecaries, and the latter "rely commonly upon the words of the silly Hearb-women, who many times bring them *Quid* for *Quo*, then which nothing can be more sad".

In this country, another strong supporter of the doctrine of signatures was the astrological botanist, Robert Turner. He definitely states that "God hath imprinted upon the Plants,

Herbs, and Flowers, as it were in Hieroglyphicks, the very signature of their Vertues".

It is satisfactory to find that the theory of signatures was repudiated by the best of the sixteenth-century herbalists. Dodoens, for instance, wrote in 1583 that "the doctrine of the Signatures of Plants has received the authority of no ancient writer who is held in any esteem: moreover it is so changeable and uncertain that, as far as science or learning is concerned, it seems absolutely unworthy of acceptance".[1]

Forty-five years later, Guy de la Brosse criticised the theory very acutely, pointing out that it was quite easy to imagine any resemblance between a plant and an animal that happened to be convenient. "C'est comme des nuées", he writes, "que l'on fait ressembler à tout ce que la fantaisie se represente, à une Gruë, à une Grenoüille, à un homme, à une armee, et autres semblables visions."

Both Paracelsus and Porta deprecate the use of foreign drugs, on the principle that, in the country where a disease arises, there nature produces means to overcome it. This idea, and variants upon it, frequently recur in herbals, being found, for example, in the writings of Tabernaemontanus, Carrichter, Culpeper, and Cole. In 1664 Robert Turner summed up these theories in the words: "For what Climate soever is subject to any particular Disease, in the same Place there grows a Cure." There is ample evidence for the survival of this theory even in the nineteenth century; for instance, in the preface to Thomas Green's *Universal Herbal* of 1816 we find the remark, "Nature has, in this country, as well as in all others, provided, in the herbs of its own growth, the remedies for the several diseases to which it is most subject." Notions of this type have been, indeed, widespread; the idea, for instance, was long current among children—and perhaps still

---

[1] "Doctrina verò de signaturis stirpium, à nullo alicuius aestimationis veterum testimonium accepit: deinde tam fluxa et incerta est, ut pro scientia aut doctrina nullatenus habenda videatur." *Pemptades*, Pempt. I, Lib. I, Cap. XI, 1583.

survives—that docks always grow near stinging-nettles, in order to provide a cure *in situ*. Whatever we may think of such beliefs, we may at least admit that the herbalists were wise to stress the use of native herbs, for, in the days of slow voyages under sail, drugs from overseas must often have become valueless by the time they reached the patient.

Paracelsus not only held the doctrine of signatures, but he also felt that, in some mystical sense, each plant was a terrestrial star, and each star was a spiritualised plant. He seems to have been conscious that

> All things by immortal power
> Near or far,
> Hiddenly
> To each other linkèd are,
> That thou canst not stir a flower
> Without troubling of a star.

Though in a more prosaic fashion, Giambattista Porta, also, believed in a connection between certain plants, on the one hand, and stars, planets, and the moon, on the other. A figure from his *Phytognomonica* here reproduced (fig. 128, p. 257) shows a number of lunar plants. To trace the history of astrology is altogether beyond our province, but we may recall an example here and there to show how universally it was accepted. We have already (p. 23) quoted the preface to the *Herbarius zu Teutsch* (1485), in which reference is made to the power and might of the goodly shining stars. Another early work, which may be mentioned in the same connection, is the *Liber aggregationis*, or *De virtutibus herbarum*, erroneously attributed to Albertus Magnus. It was first printed in the fifteenth century, and translated into many languages, one of the English versions appearing about 1565 under the title: *The boke of secretes of Albartus Magnus, of the vertues of Herbes, stones and certaine beastes*. This volume does not contain much information about plants, being mostly occupied with animals and minerals, but such "botany" as it includes is fully astrological. For instance we are told that if the marigold "be gathered, the Sunne beynge in the sygne Leo, in August, and be wrapped in the leafe of a Laurell, or baye

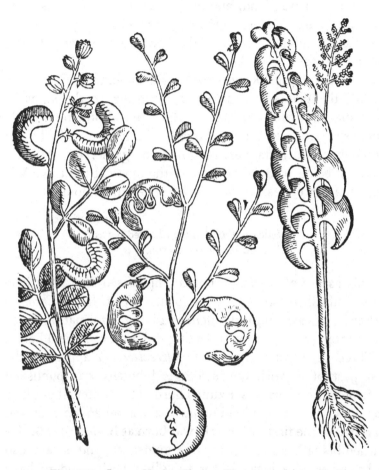

Fig. 128. Lunar Herbs [Porta, *Phytognomonica*, 1591]

tree, and a wolves tothe be added therto, no man shalbe able to have a word to speake agaynst the bearer therof, but woordes of peace". Concerning the plantain we read, "The rote of this herbe is mervalous good agaynst the payne of the headde, because the signe of the Ramme is supposed to be the house of the planete Mars, which is the head of the whole worlde."

Astrological ideas must, indeed, have been very familiar in Elizabethan England, and are reflected in many passages in Shakespeare, never, perhaps, more charmingly than in Beatrice's laughing words—"there was a star danced, and under that was I born".

In the astrological *Kreutterbůch* of Bartholomaeus Carrichter (1575), the plants are arranged according to the signs of the zodiac, but the principle on which this allotment is made remains mysterious to the lay mind. Much stress is laid upon the hour at which the herbs ought to be gathered, particular attention being paid to the state of the moon at the time. We are reminded of Jessica, who says, under the moonlight:

> In such a night,
> Medea gather'd the enchanted herbs
> That did renew old Aeson.

This aspect of the subject was emphasised also in a little book published in 1571, Nicolaus Winckler's *Chronica herbarum*, an astrological calendar giving information as to the appropriate times for collecting different roots and herbs.

Three years after Carrichter's *Kreutterbůch* appeared, the first part of a work on astrological botany was published by Leonhardt Thurneisser zum Thurn. This writer, though he was possessed of genuine talent, was also an adventurer and charlatan of the first order. He was born at Basle in 1530. He learned his father's craft, that of a goldsmith, and is said also to have helped a local doctor to collect herbs, and to have been employed to read aloud to him from the works of Paracelsus. His career in Basle came to an untimely end, for, finding himself in financial straits, he tried to retrieve his position by substituting gilded lead for gold. The fraud was discovered, and he had to flee the country. He travelled widely, making an especial study of mining. He had an adventurous and varied life, with alternations of poverty and wealth.

During Thurneisser's most influential period, he lived in Berlin, practising medicine, and making amulets, talismans, and secret remedies which yielded large profits. To the rich he commended and sold such costly drugs as gold drops, tincture of pearls, and water of amethysts. He also published astrological calendars, and cast nativities. He owned a printing press, and employed a large staff, which included artists and engravers. Eventually he fell upon evil days, and he is said to have died in obscurity in the last decade of the sixteenth century.

In the history of botany, Thurneisser has his little place, since he formed a cabinet of natural curiosities, including a large collection of dried plants and of seeds, and he cultivated many rare herbs. He projected also the vast work on astrological botany to which we have alluded. This was intended to occupy ten books, but, though the first was published in Berlin in 1578, the remaining volumes never saw the light. The title was *Historia unnd Beschreibung Influentischer, Elementischer und Natürlicher Wirckungen, Aller fremden unnd heimischen Erdgewechssen.* A Latin version of this book, under the name *Historia sive descriptio plantarum,* was published in the same year. This first instalment deals only with the umbellifers, which were regarded as under the dominion of the Sun and Mars. Neither nomenclature nor figures are clear enough to allow individual species to be recognised. Each plant is drawn in an ellipse, surrounded by an ornamental border, which contains mysterious inscriptions relating to its virtues (cf. pl. xxv, f.p. 260). In some cases diagrams are added, which are possibly horoscope charts, giving indications as to the course of an illness, and as to the appropriate herbal treatment (e.g. fig. 129, p. 260).

After the manner of the ancients, Thurneisser describes plants, according to their degree of vigour, as either male or female. He also adds a third class, typified by a child, to symbolise those whose qualities are feeble. It may perhaps be worth while to translate here a few sentences of the first

CONSTELLATIO PECVLIARIS.

PLANTA hac anno 1528 collecta eft, a D, Friderico Stren-
gero Medico & Phyfico excellenti: & prater Vires commemoratas
etiam alijs multis pollere deprehenfa eff. Tempore autem foffionis
illius, Calum pofitu Planetarum & Signorum (Vt Figura appofita
oftendit) talem fere faciem obtinuit,

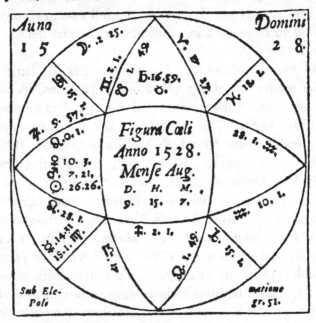

Fig. 129. Astrological Diagram for "Cervaria faemina" [Thurneisser, *Historia sive descriptio plantarum*, 1587]

chapter of the *Historia*,[1] to show how far such writers as Leonhardt Thurneisser had departed from the study of botany on rational lines. When discussing the planting of roots and herbs and the gathering of seeds, he declares that "it is absolutely essential that these operations should be performed

---

[1] The edition of 1587 was used in making this translation.

Plate x.xv

"Cervaria faemina" [Thurneisser, *Historia sive descriptio plantarum*, 1587]

so as to correspond with the stations and positions of the planets and heavenly bodies, to whose control diseases are properly subject. And against disease we have to employ herbs, with due regard of course to the sex, whichever it be, of human beings; and so herbs intended to benefit the male sex should be procured when the Sun or Moon is in some male sign [of the Zodiac], e.g. Sagittarius or Aquarius, or if this is impossible, at least when they are in Leo. Similarly herbs intended to benefit women should be gathered under some female sign, Virgo, of course, or, if that is impossible, in Taurus or Cancer".

In the seventeenth century, England became badly infected with astrological botany. The most notorious exponent of the subject was Nicholas Culpeper (1616–54), who, about 1640, set up as an astrologer and physician in Spitalfields. His portrait is reproduced in pl. xxvi, f.p. 262. He aroused great indignation among the medical profession by publishing, under the name of *A Physicall Directory*, an unauthorised English translation of the *Pharmacopoeia*, which had been issued by the College of Physicians. That Culpeper was unpopular with orthodox medical practitioners is hardly surprising, when we consider the way in which he speaks of them in this book, as "a company of proud, insulting, domineering Doctors, whose wits were born above five hundred years before themselves". He goes on to ask—"Is it handsom and wel-beseeming a Common-wealth to see a Doctor ride in State, in Plush with a footcloath, and not a grain of Wit but what was in print before he was born?"

Many editions of the *Physicall Directory* were issued under different names. That of 1661 is described on the title-page as *Being an Astrologo-Phisical Discourse of the Vulgar Herbs of this Nation: Containing a Compleat Method of Physick, whereby a man may preserve his Body in Health; or Cure himself, being Sick, for three pence Charge, with such things only as grow in* England, *they being most fit for English Bodies.*

VIII. Signatures and Astrology

Culpeper describes certain herbs as being under the dominion of the sun, the moon, or a planet, and others as under a planet and also one of the constellations of the Zodiac. His reasons for connecting a particular herb with a particular heavenly body could scarcely be more inconsequent. He states, for example, that "Wormwood is an Herb of *Mars,* . . . I prove it thus: What delights in Martial places, is a Martial Herb; but Wormwood delights in Martial places (for about Forges and Iron Works you may gather a Cart load of it) *Ergo* it is a Martial Herb".

Since each disease, on Culpeper's view, is caused by a planet, one way of curing the ailment is by the use of herbs belonging to an opposing planet—e.g. diseases produced by Jupiter are healed by the herbs of Mercury. On the other hand, the illness may be cured "by sympathy", that is by the use of herbs belonging to the planet which is responsible for the disease. As an example we may quote his explanation of the use of wormwood for affections of the eyes. "The eyes are under the Luminaries; the right eye of a man, and the left eye of a woman the *Sun* claims Dominion over: The left eye of a man, and the right eye of a woman, are the priviledg of the *Moon,* Wormwood an herb of *Mars* cures both what belongs to the *Sun* by Sympathy, because he is exalted in his house; but what belong to the *Moon* by Antipathy, because he hath his Fall in hers."

It is amusing to find that, in his preface, Culpeper claims that he surpasses all his predecessors in being alone guided by reason, whereas all previous writers are "as ful of nonsense and contradictions as an Eg is ful of meat".

Even in the seventeenth century, Culpeper was severely criticised. William Cole, despite his credulity about the doctrine of signatures, was whole-hearted in his condemnation of astrological botany. The main argument, by which he sought to discredit it, was a peculiarly ingenious one. The knowledge of herbs is, he says, "a subject as antient as the

Plate xxvi

In Effigiem Nicholai Culpeper Equitis .

The shaddow of that Body heer you find
Which serves but as a case to hold his mind,
His Intellectuall part be pleas'd to looke
In lively lines described in the Booke . *Crosse sculpsit*

NICHOLAS CULPEPER (1616–1654)
[*A Physicall Directory*, 1649. Engraving by Thomas Cross]

Nicholas Culpeper

Creation (as the Scriptures witnesse) yea more antient then the Sunne, or Moon, or Starres, they being created on the fourth day, whereas Plants were the third. Thus did God even at first confute the folly of those Astrologers, who goe about to maintaine that all vegitables in their growth, are enslaved to a necessary and unavoidable dependance on the influences of the Starres; Whereas Plants were, even when Planets were not".

Cole not only attacked astrological botany in general, but he particularly condemned "Master *Culpeper* (a man now dead, and therefore I shall speak of him as modestly as I can, for were he alive, I should be more plain with him)...;...he, forsooth, judgeth all men unfit to be Physitians, who are not Artists in Astrology, as if he and some other Figure-flingers his companions, had been the onely Physitians in *England*, whereas for ought I can gather, either by his Books, or learne from the report of others, he was a man very ignorant in the forme of Simples".

Cole's opposition had little effect; Culpeper's herbal ran into edition after edition, and even to-day—not much less than three hundred years after its first appearance—it still finds an appreciative public. A version was published as recently as 1932; as an earnest of its quality, it will suffice to quote a few lines, chosen from among many others of the same type: "Mushrooms are under the dominion of Saturn and if any are poisoned by eating them, Wormwood as an herb of Mars cures him, because Mars is exalted in Capricorn, the House of Saturn; and this is done by sympathy."

Chapter IX

CONCLUSIONS

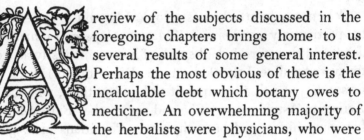 review of the subjects discussed in the foregoing chapters brings home to us several results of some general interest. Perhaps the most obvious of these is the incalculable debt which botany owes to medicine. An overwhelming majority of the herbalists were physicians, who were led to the study of botany on account of its connection with the arts of healing. As we have already pointed out, medicine gave the original impulse, not only to systematic botany, but also to the study of the anatomy of plants.

However, as the evolution of the herbal proceeded, we have shown that botany rose from being a mere handmaid of medicine to a position of comparative independence. This is well exemplified in the history of plant classification. When the early medical botanists attempted any distribution of their material, it was on a purely utilitarian basis. The herbs were at first arranged merely according to the qualities which made them of value to man; but, as the science grew, the need for a classification of a different type began to make itself felt, and in some of the works published in the latter half of the period which we have been considering, there is a distinct, if only partially successful, attempt towards grouping plants according to the affinities which they present when considered in themselves, and not in relation to man. The ideal of a natural system in the vegetable kingdom, in which each kind should find its inevitable place, must have been clear to

The Rise of Botany

Mathias de l'Obel, when he wrote, in the *Adversaria*, of "an order, than which nothing more beautiful exists in the heavens, or in the mind of a wise man".[1]

Second only to the debt of botany to medicine, is its debt to certain branches of the arts, especially printing and wood-engraving. Through these two associated crafts, the traditional lore recorded in the manuscript herbals was embodied, with little loss of continuity, in the printed herbals which succeeded them. The draughtsman and engraver, indeed, did more than merely disseminate existing knowledge; they lent the botanist their acute and highly trained powers of observation, and their work must often have revealed to him much that, without their help, he would never have seen.

The art of plant description, in its historical development, lagged conspicuously behind that of plant illustration. The picturesque but vague and crude phytography, characteristic of the early herbals, had to pass through a long period of refining evolution before it could become an instrument of scientific precision.

The rapid rise of botany in the two centuries which we have reviewed, must have been greatly stimulated by the cosmopolitanism of the savants of the renaissance. Study at more than one foreign university, aided by the use of the Latin tongue as a *lingua franca*, formed part of the recognised routine. It is on record that, when the city fathers of Zurich in 1533 sent Konrad Gesner to study in France, it was with the deliberate intention of saving him from "having seen Zurich alone and heard Zurich alone". Despite the difficulty and danger of travelling in those days, the botanists of the sixteenth and seventeenth centuries vied with those of the twentieth in their extensive knowledge of Europe, and in the intimacy of their relations with men of science in other countries. In some cases this zeal for travel was not altogether spontaneous, but was fostered by the religious disturbances,

[1] "Sic enim ordine, quo nihil pulchrius in coelo, aut in Sapientis animo,..."

at the period of the Reformation and later, which so often drove into exile those who could no longer accept the authority of the Church of Rome. William Turner, Charles de l'Écluse, and the Bauhins, suffered this enforced nomadism. The number of botanists, in central and northern Europe, who were of the reformed faith, is, indeed, remarkable. Other Protestants, in addition to those who have just been cited, were Brunfels, Bock, Fuchs, Tabernaemontanus, and Boldizsár de Batthyány; with these names we may also associate that of Christophe Plantin, the publisher who did so much for botany, for, though he was a Catholic by profession, he was a secret adherent of the Familists, the precursors of the Quakers.

The same independence of thought, which led many botanists to throw in their lot with the spiritual reformers of their day, also led them, within their own special field, to discard many of the irrational beliefs which were then rife. In the works of those herbalists of the sixteenth and seventeenth centuries who are in the first rank—Bock, William Turner, Dodoens, Gaspard Bauhin, and others—we find little approval of any kind of superstition connected with plants, such as the doctrine of signatures, or astrological speculations. A number of books dealing with topics of this type appeared during the period which we have studied, but their writers form a class apart, and must not be confused with the herbalists proper, whose attitude was, on the whole, marked by a healthy scepticism, in which they were in advance of their time. It would be far from true to say that they were all quite free from superstition, but it was well controlled, and magical short cuts to health were not commended, even if not entirely disbelieved. Laurence Andrew, the translator of Jerome of Brunswick's *Boke of Distyllacyon*, wrote in 1527, "Beholde how moche it excecedeth to use medicyme of eficacye naturall by god ordeyned then wycked wordes or charmes of efycacie unnaturall by the devyll envented."

The Printed Herbal

Despite the technical nature of their subject, the personality of the herbalists is often revealed between the lines with surprising clearness. We may say of many of the early botanical books, as William Turner said of his own—"thys worke...Which though it be but lyttle, yet it is able to declare my mynde thorowly, as yᵉ lyones clawe only sene, bewrayeth the hole lyone".

Fig. 130. Woodcut from Title-page
[*The grete herball*, 1526] *Reduced*

When we cease to analyse our material, and attempt to synthesise the preceding chapters, we see that our concern has been with the most active life-period of the printed herbal, which may be reckoned as beginning in the last quarter of the fifteenth century, with *The Book of Nature*, the *Herbarium* of Apuleius Platonicus, and the *Latin* and the *German Herbarius*. When this active period should be said to end, is less easily decided, but in some senses it may fairly be

267

IX. Conclusions

taken as concluding round about the year 1670, and thus as covering only the comparatively short space of two hundred years. From the point of view of illustration, the best epoch in the history of the printed herbal is confined within much narrower limits than these two centuries. The suggestion has been made, and seems to be thoroughly justified, that the finest period should be reckoned as falling between 1530 and 1614, that is, between the woodcuts after Hans Weiditz in Brunfels' *Herbarum vivae eicones*, and the copperplates of Crispian de Passe in the *Hortus floridus*. This period may be held to culminate, artistically, in the woodcuts in Fuchs' herbals, and, scientifically, in those of Gesner-Camerarius.

As far as the text is concerned, the meridian of the botanical works in our special period is, as it were, dimly foreseen in the *Stirpium adversaria nova* of Pena and de l'Obel (1570,1), and attained, fifty years later, in the *Prodromos* (1620) and the *Pinax* (1623) of Gaspard Bauhin. In the works of Bauhin, classification, nomenclature, and description reach their high-water mark, though it is to de l'Obel, and to his precursor, Bock, who was also a pioneer in phytography, that we owe the first definite efforts after a natural system.

In the latter part of the seventeenth century, botany rapidly became more scientific; the discovery of the function of the stamens, which was first announced in 1682, marked a very definite step in advance. As time went on, the *herbal*, with its characteristic mixture of medical and botanical lore, gave way before the exclusively medical *pharmacopoeia* on the one hand, and the exclusively botanical *flora* on the other. As the use of home-made remedies declined, and the chemist's shop took the place of the housewife's herb-garden and still-room, the practical value of the herbal diminished almost to vanishing point.

If, lifting our eyes from the relatively limited period chosen for special study in this book, we look before and after, and try to visualise the history of the herbal as a whole, we

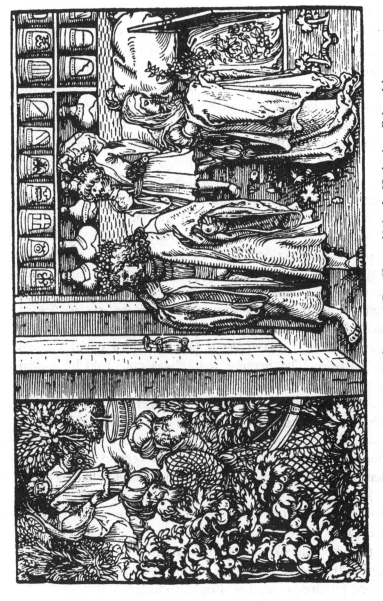

Fig. 131. Woodcut from Title-page [*Das Kreüterbüch oder Herbarius.* Printed by Heinrich Stayner, Augsburg, 1534]

IX. Conclusions

cannot but be struck by its *continuity*. The earliest phases of its evolution are lost to us, but we can at least see that, like most of the things of the mind, it can be traced back to the thought of ancient Greece; and from a time before the opening of the Christian era, it can be followed lineally to our own day. There were epochs in which its history was merely a tale of marking time, or back-sliding, but, even in the darkest ages, scribes among the Syriac- and Arabic-speaking peoples copied and translated classical manuscripts, and kept their contents from falling into oblivion. Access to Greek texts at the time of the revival of learning, renewed the study of plants in western Europe. The value of the classical tradition was great even for those who brought their observations to it as to an infallible touchstone, but it was far greater for those who found in it a stimulus to unending critical discussion. Out of the interaction of these two attitudes, systematic botany, as we know it to-day, has gradually developed. Sibthorp's monumental *Flora Graeca* is, indeed, the direct descendant in modern science of the *De materia medica* of Dioscorides. Greek influence has not only permeated scholarly work of this type; it has also lived on in a more homely world. The present writer was once told by a man who was born in 1842 that, during his boyhood in Bedfordshire, he was acquainted with a cottager who treated the ailments of her neighbours with the help of a copy of Gerard's *Herball*. If, as is most likely, this was one of Johnson's editions, she must thence have known certain illustrations copied from Anicia Juliana's manuscript of Dioscorides made soon after A.D. 500—figures which were probably themselves derived from the work of Krateuas, belonging to the century before Christ. We thus catch a glimpse of the herbal tradition passing unbroken through two thousand years, from Krateuas, the Greek, to an old woman poring over her well-thumbed picture-book in an English village.

APPENDIX I

A CHRONOLOGICAL LIST OF THE PRINCIPAL HERBALS AND RELATED BOTANICAL WORKS PUBLISHED BETWEEN 1470 AND 1670

[This list, which is intended for the botanist rather than the bibliographer, is far from being exhaustive, especially as regards works published in the seventeenth century. In most cases reference is made to first editions only. Subsequent editions and translations, though often numerous and important, are usually not cited unless special mention has been made of them in the text. In cases where such editions are quoted, their titles are placed beneath that of the first edition (i.e. under the date of the first edition). Independent works by the same author are, however, arranged chronologically, so that, in this list, all the works of any given author are not placed together, but must be looked for under their respective dates; the titles in this Appendix are included in the general Index. The author's name, or, in the case of anonymous works, the title most commonly used, is printed in heavy type. All the works enumerated have been examined personally by the writer, except two, to which footnotes are appended.]

? 1472

Bartholomaeus Anglicus [erroneously called Glanville, Bartholomew de]. Liber de proprietatibus rerum. *Begins:* Incipit prohemium de proprietatibus rerum fratris bartholomei anglici de ordine fratrum minorum. [?Cologne, ?1472.] [*A general work containing one section dealing with plants.*]

—— (*Another Edition*). Liber de proprietatibus rerum. [?Westminster, ?1495.] [*A translation by Trevisa printed by Wynkyn de Worde.*]

1475

Konrad von Megenberg [Cûnrat]. *Begins:* Hÿe nach volget das pûch der natur.... Hanns Bâmler. Augspurg, 1475. [*A general work containing a section dealing with plants.*]

271

Appendix I

1477

Macer, Aemilius [Odo]. *Begins:* Incipit liber Macri philosophi in quo tractat de naturis qualitatibus et virtutibus Octuagintaocto herbarum. (*Colophon:*) Neapoli impressus per Arnoldum de Bruxella. 1477.[1]

—— (*Another Edition*). Herbarum varias q̇ vis cognoscere vires Macer adest: disce quo duce doct' eris. *Begins:* Macer floridus de viribus herbarum. (*Colophon:*) Impressus Parisius per Magistrum Johannem Seurre. Pro Magistro Petro Bacquelier. 1506.

—— (*Another Edition*). Les fleurs du livre des vertus des herbes, composé iadis en vers Latins par Macer Floride:...Le tout mis en François, par M. Lucas Tremblay, Parisien.... Rouen. Martin, et Honoré Mallard. 1588.

1478

Dioscorides. *Begins:* [M]ulti voluerunt (*Marginal glosses begin:* Notādum...diascorides). (*Colophon:*) Impressus colle...ioh'em allemanum de Medemblick. 1478. [*The first edition of* De materia medica *in Latin.*]

—— (*Another Edition*). περὶ ὕλης ἰατρικῆς. Aldus Manutius, Venice, 1499. [*The first edition of* De materia medica *in Greek.*]

Albertus Magnus [erroneously attributed to]. Liber aggregationis seu liber secretorum Alberti magni de virtutibus herbarum... (*Colophon:*) per Magistrum Johannem de Annunciata de Augusta. 1478.[2]

—— (*Another Edition*). De virtutibus herbarum. De virtutibus lapidum De virtutibus animalium et mirabilibus mundi. Thomas Laisne, Rouen. [? 1500.]

—— (*Another Edition*). The boke of secretes of Albartus Magnus, of the vertues of Herbes, stones and certaine beastes. Also a boke of the same author, of the marvaylous thinges of the world.... London. Wyllyam Copland. [? 1560.]

? 1481[3]

Apuleius Platonicus. *Begins:* Incipit Herbarium Apulei Platonici ad Marcum Agrippam. J. P. de Lignamine. [Rome, ? 1481.]

1483

Theophrastus. De causis plantarum lib. vi. Impressus Tarvisii per Bartholomaeum Confalonerium de Salodio. 1483. [*First Latin edition.*]

—— (*Another Edition*). περὶ φυτῶν. Aldus Manutius, Venice, 1497. [*The first edition in Greek, included in the Aldine Aristotle.*]

[1] The writer has examined a facsimile rotograph of this extremely rare book, from the copy in the John Rylands Library, Manchester.

[2] This is probably not the first edition.

[3] The *Gesamtkatalog der Wiegendrucke* gives the date as 1483/4, while Hunger, F. W. T. (1935[2]) gives 1481.

Appendix I

1484

The Latin Herbarius [referred to by various authors as Herbarius in Latino, Aggregator de Simplicibus, Herbarius Moguntinus, Herbarius Patavinus, etc.]. Herbarius Maguntie impressus. [Peter Schöffer. Mainz.] 1484.

—— (*Another Edition*). *Begins:* Dye prologhe de oversetters uyt den latyn in dyetsche. [Veldener, Kuilenburg.] 1484. [*A Flemish translation.*]

—— (*Another Edition*). *Begins:* Incipit Tractatus de virtutibus herbarum. (*Colophon:*) Impressum Venetiis per Simonem Papiensem dictum Biuilaquam.... 1499. [Sometimes called "Herbarius Arnoldi de nova villa Avicenna".]

1485

The German Herbarius [referred to by various authors as the Herbarius zu Teutsch, Gart der Gesundtheit, the German Ortus Sanitatis, the smaller Ortus, Johann von Cube's Herbal, etc.]. *Begins:* Offt und vil habe ich. [Peter Schöffer.] Mencz, 1485.

.—— (*Another Edition*). *Begins:* Offt und vil hab ich. [Schönsperger.] Augspurg, 1485.

1491

Ortus Sanitatis [Hortus Sanitatis]. *Prohemium begins:* Omnipotentis eternique dei.... (*Colophon:*) Jacobus Meydenbach. Moguntia, 1491.

—— (*Another Edition*). (*Colophon:*) Impressum Venetiis per Bernardinum Benalium: Et Joannem de Cereto de Tridino alias Tacuinum. 1511.

—— (*Another Edition*). Ortus sanitatis translate de latin en francois. Anthoine Verard. Paris, n.d. [?1501].

—— (*Another Edition*). Le jardin de sante translate de latin en francoys nouvellement Imprime a Paris. On les vend a Paris en la rue sainct Jacques a lenseigne de la Rose blanche couronnee. (*Colophon:*) Imprime a Paris par Philippe le noir. [?1539.]

1500

Braunschweig, Hieronymus [Jerome of Brunswick]. Liber de arte distillandi. de Simplicibus. Johannes Grüeninger. Strassburg, 1500.

—— (*Another Edition*). The vertuose boke of Distyllacyon of the waters of all maner of Herbes....Laurens Andrewe. London, 1527.

1516

Ruellius, Johannes [Ruel, Jean]. Pedacii Dioscoridis Anazarbei de medicinali materia libri quinque....Impressum est in...Parrhisiorum Gymnasio...in officina Henrici Stephani. 1516.

Appendix I

1525

Herball. Here begynneth a newe mater, the whiche sheweth and treateth of ẏ vertues and proprytes of herbes, the whiche is called an Herball. Rycharde Banckes. London, 1525.

—— (*Another Edition*). Macers Herbal. Practysyd by Docter Lynacro. Robert Wyer. n.d. [London, ? 1530].

—— (*Another Edition*). A new Herball of Macer, Translated out of Laten in to Englysshe. Robert Wyer...in seynt Martyns Parysshe... besyde Charynge Crosse. London, n.d. [? 1535].

—— (*Another Edition*). A boke of the propreties of Herbes called an herball, wherunto is added the tyme ẏ herbes, Floures and Sedes shoulde be gathered...by W. C.[1] Wyllyam Copland. London, n.d. [1550].

—— (*Another Edition*). A litle Herball of the properties of Herbes... wyth certayne Additions at the ende of the boke, declaring what Herbes hath influence of certain Sterres...Anthony Askham, Physycyon. Jhon Kynge. London, n.d. [1555 or later].

Before 1526

Grand Herbier. Le grand Herbier en Francoys: contenant les qualitez, vertus et proprietez des herbes, arbres, gommes....Pierre Sergent. Paris, n.d.

—— (*Another Edition*). The grete herball whiche geveth parfyt knowlege and understandyng of all maner of herbes and there gracyous vertues.... (*Colophon:*) Peter Treveris. London, 1526.

—— (*Another Edition*). The grete herball.... (*Colophon:*) Imprynted at London...by me Peter Treveris.... 1529.

1530

Brunfelsius, Otho [Brunfels, Otto von]. Herbarum vivae eicones... Argentorati apud Joannem Schottum. 1530, 1, 6. (*Another Edition*). 1532,6.

—— (*Other Editions*). Contrafayt Kreüterbůch. Strasszburg. Schotten. 1532,7. Kreüterbůch contrafayt. Strasszburg. Schotten. 1534.

1533

Rhodion, Eucharius [Rösslin]. Kreutterbůch...Anfenglich von Doctor Johan Cuba zusamen bracht, Jetz widerum new Corrigirt...Mit warer Abconterfeitung aller Kreuter. Zu Franckfurt am Meyn, Bei Christian Egenolph. 1533. [*A large number of editions of this work appeared, edited by Dorstenius, Lonicerus, and others.*]

[1] The initials "W. C." probably refer to William Copland, and not, as has sometimes been supposed, to Walter Cary; see Barlow, Appendix II.

Appendix I

1536

Amatus Lusitanus [Castell-Branco, J. R. de]. Index Dioscoridis....
Excudebat Antverpiae vidua Martini Caesaris. *1536*.

Ruellius, Joannes [Ruel, Jean]. De Natura stirpium libri tres,...
Parisiis Ex officina Simonis Colinaei. *1536*.

1537

Brasavola, Antonio Musa. Examen omnium Simplicium....
Lugduni,...apud Ioannem et Franciscum Frellaeos, Fratres. *1537*.[1]

1538

Turner, William. Libellus de re herbaria novus, in quo herbarum
aliquot nomina greca, latina, et Anglica habes, una cum nominibus
officinarum.... Londini apud Ioannem Byddellum. *1538*.

1539

Tragus, Hieronymus [Bock, Jerome]. New Kreütter Bûch von
underscheydt, würckung und namen...gedruckt zů Strassburg, durch
Wendel Rihel. *1539*.

—— (*Other Editions*). Kreuter Bûch. Wendel Rihel. Strasburg, *1546*;
Strassburg, *1551*.

—— (*Another Edition*). De stirpium, maxime earum, quae in Ger-
mania nostra nascuntur...nunc in Latinam conversi, Interprete Davide
Kybero.... (*Colophon:*) Argentorati Excudebat Vuendelinus Rihelius....
1552.

1541

Gesnerus, Conradus [Gesner, Konrad]. Historia plantarum et
vires ex Dioscoride, Paulo Aegineta,... Parisiis Apud Ioannem Lodoicum
Tiletanum. *1541*.

1542

Fuchsius, Leonhardus [Fuchs, Leonhart]. De historia stirpium....
Basileae, in officina Isingriniana.... *1542*.

—— (*Another Edition*). New Kreüterbûch. Michael Isingrin. Basell,
1543.

—— (*Another Edition*). Leonharti Fuchsii medici, primi de stirpium
historia commentariorum tomi vivae imagines, in exiguam...formam
contractae....Isingrin. Basileae, *1545*.

---

[1] An edition of 1536 is referred to, but it has not been seen by the writer.

Appendix I

Gesnerus, Conradus [Gesner, Konrad]. Catalogus plantarum Latinè, Graecè, Germanicè, et Gallicè.... Tiguri apud Christoph. Froschouerum. 1542.

1544

Matthiolus, Petrus Andreas [Mattioli, Pierandrea]. Di Pedacio Dioscoride Anazarbeo libri cinque Della historia, et materia medicinale tradotti in lingua volgare Italiana.... Venetia per Nicolo de Bascarini da Pavone di Brescia. 1544.

—— Commentarii, in libros sex Pedacii Dioscoridis Anazarbei, de medica materia.... Venetiis...apud Vincentium Valgrisium. 1554.

—— (*Another Edition*). Herbarz: ginak Bylinář,... [Prague, 1562.]

—— (*Another Edition*). Commentarii in sex libros Pedacii Dioscoridis Anazarbei de Medica materia,... Venetiis, Ex Officina Valgrisiana. 1565.

1548

Turner, William. The names of herbes in Greke, Latin, Englishe Duche and Frenche wyth the commune names that Herbaries and Apotecaries use. John Day and Wyllyam Seres. London, 1548.

1551

Turner, William. A new Herball. Steven Mierdman. London, 1551.

—— The seconde parte of Vuilliam Turners herball. Arnold Birckman. Collen, 1562.

—— The first and seconde partes of the Herbal of William Turner... with the Third parte, lately gathered.... Arnold Birckman. Collen, 1568.

1553

Amatus Lusitanus [Castell-Branco, J. R. de]. In Dioscoridis Anazarbei de medica materia libros quinque enarrationes.... Venetiis, 1553. (*Colophon:*) apud Gualterum scotum.

Bellonius, Petrus [Belon, Pierre]. De arboribus coniferis, resiniferis, aliis quoque nonnullis sempiterna fronde virentibus,... Parisiis Apud Gulielmum Cavellat,... 1553.

276

Appendix I

1554

Dodonaeus, Rembertus [Dodoens, Rembert]. Crǔẏdeboeck. (*Colophon:*) Ghedruckt Tantwerpen...Jan vander Loe.... 1554.
—— (*Another Edition*). Histoire des plantes,...Nouvellement traduite...en François par Charles de l'Escluse. Jean Loë. Anvers, 1557.
—— (*Another Edition*). A Nievve Herball, or Historie of Plantes:.... nowe first translated out of French into English, by Henry Lyte Esquyer. At London by me Gerard Dewes.... 1578. (*Colophon:*) Imprinted at Antwerpe, by me Henry Loë Bookeprinter, and are to be solde at London in Povvels Churchyarde, by Gerard Devves.[1]

1559

Maranta, Bartholomaeus. Methodi cognoscendorum simplicium libri tres. Venetiis Ex officina Erasmiana Vincentii Valgrisii. 1559.

1561

Anguillara, Luigi. Semplici dell' eccellente M.L.A., Liquali in piu Pareri à diversi nobili huomini scritti appaiono, Et Nuovamente da M. Giovanni Marinello mandati in luce. In Vinegia...Vincenzo Valgrisi, 1561.

Cordus, Valerius. In hoc volumine continentur Valerii Cordi... Annotationes in Pedacii Dioscoridis...de Medica materia...eiusdem Val. Cordi historiae stirpium lib. IIII....Omnia...Conr. Gesneri... collecta, et praefationibus illustrata. (*Colophon:*) Argentorati excudebat Iosias Rihelius. 1561.

1563

Horto, Garcia ab [de Orta]. Coloquios dos simples, e drogas he cousas mediçinais da India.... Impresso em Goa, por Ioannes de endem as x. dias de Abril de 1563. annos.

1564

Mizaldus, Antonius [Mizauld, Antoine]. Alexikepus, seu auxiliaris hortus,... Lutetiae, Apud Federicum Morellum.... 1564.

1566

Dodonaeus, Rembertus [Dodoens, Rembert]. Frumentorum, leguminum, palustrium et aquatilium herbarum...historia:... Antverpiae, Ex officina Christophori Plantini. 1566.

[1] This edition is also found with a variant title-page, in which the reference to Dewes is replaced by the statement, "Imprinted at Antvverpe, by me Henry Loë".

277

Appendix I

1568

Dodonaeus, Rembertus [Dodoens, Rembert]. Florum, et coronariarum odoratarumque nonnullarum herbarum historia,.... Antverpiae, Ex officina Christophori Plantini. 1568.

1569

Monardes, Nicolas. Dos libros, el veno que trata de todas las cosas que traen de nuestras Indias Occidentales.... Impressos en Sevilla en casa de Hernando Diaz.... 1569.

—— Segunda parte del libro, de las cosas que se traen de nuestras Indias Occidentales.... En sevilla En casa Alonso Escrivano. 1571.

—— Primera y segunda y tercera partes de la historia medicinal de las cosas que se traen de nuestras Indias Occidentales.... En Sevilla. 1574.

—— (*Another Edition*). Joyfull newes out of the newe founde worlde, wherein is declared the rare and singuler vertues of diverse...Hearbes.... Englished by Jhon Frampton Marchaunt. London, W. Norton, 1577. [*This book was also issued with the same printer and date, but with another title-page:* The Three Bookes written in the Spanishe tonge, by the famous Phisition D. Monardes...translated into Englishe by Jhon Frampton Marchant.]

1570

Paracelsus [Bombast von Hohenheim]. Ettliche Tractatus des hocherfarnen unnd berümbtesten Philippi Theophrasti Paracelsi.... I. Von Natürlichen dingen. II. Beschreibung etilcher kreütter. III. Von Metallen. IV. Von Mineralen. V. Von Edlen Gesteinen. Strassburg. Christian Müllers Erben. 1570.

—— Paracelsus his Dispensatory and Chirurgery....Faithfully Englished, by W. D. London: Printed by T. M. for Philip Chetwind.... 1656.

1570, 1

Lobelius, Mathias [de l'Obel or de Lobel, Mathias] and **Pena, Petrus** [Pena, Pierre]. Stirpium adversaria nova. Londini. 1570. (*Colophon:*) Londini. 1571. ...excudebat prelum Thomae Purfoetii.

—— (*Another Edition*). Nova stirpium adversaria,... Antverpiae Apud Christophorum Plantinum. 1576.

Lobelius, Mathias. Plantarum seu stirpium historia,...Cui annexum est Adversariorum volumen. Antverpiae, Ex officina Christophori Plantini. 1576.

—— (*Another Edition*). Kruydtboeck. T'Antwerpen. By Christoffel Plantyn. 1581.

Appendix I

1571

Matthiolus, Petrus Andreas [Mattioli, Pierandrea]. Compendium De Plantis omnibus,...de quibus scripsit suis in commentariis in Dioscoridem editis.... Accessit praeterea ad calcem Opusculum de itinere, quo è Verona in Baldum montem Plantarum refertissimum itur... Francisco Calceolario...Venetiis, In Officina Valgrisiana. 1571.

Winckler, Nicolaus. Chronica herbarum, florum, seminum,... Augustae Vindelicorum, in officina Typographica Michaëlis Mangeri. 1571.

1574

Dodonaeus, Rembertus [Dodoens, Rembert]. Purgantium aliarumque eo facientium, tum et Radicum, Convolvulorum ac deleteriarum herbarum historiae libri iiii. Antverpiae, Ex officina Christophori Plantini. 1574.

1575

Carrichter, Bartholomaeus. Kreutterbûch.... Gedruckt zû Strassburg...bey Christian Müller. 1575.

1576

Clusius, Carolus [l'Écluse or l'Escluse, Charles de]. Caroli Clusii Atrebat. Rariorum aliquot stirpium per Hispanias observatarum Historia, ... Antverpiae, Ex officina Christophori Plantini,... 1576.

1578

Acosta, Christoval. Tractado de las drogas y medicinas de las Indias Orientales con sus Plantas. En Burgos. Por Martin de Victoria.... 1578.

Thurneisserus, Leonhardus [Thurneisser zum Thurn, Leonhardt]. Historia sive descriptio plantarum.... (*Colophon:*) Berlini Excudebat Michael Hentzske. 1578.

—— (*Another Edition*). Historia unnd Beschreibung Influentischer, Elementischer und Natürlicher Wirckungen, Aller fremden unnd Heimischen Erdgewechssen.... (*Colophon:*) Gedruckt zu Berlin, bey Michael Hentzsken. 1578.

—— (*Another Edition*). Historia sive descriptio plantarum.... Coloniae Agrippinae, apud Ioannem Gymnicum,... 1587.

1580

Dodonaeus, Rembertus [Dodoens, Rembert]. Historia vitis vinique: et stirpium nonnullarum aliarum. Coloniae Apud Maternum Cholinum. 1580.

Appendix I

1581

Lobelius, Mathias [de l'Obel or de Lobel, Mathias]. Plantarum seu stirpium icones. Antverpiae, Ex officina Christophori Plantini. 1581.

1582, 3

Rauwolff, Leonhard. Leonharti Rauwolfen,... Aigentliche beschreibung der Raiss, so er vor diser zeit gegen Auffgang inn die Morgenländer,... 1582, 1583. [*This is a book of travel, but the fourth part, which has a separate title-page, dated 1583, Getruckt zů Laugingen, durch Leonhart Reinmichel, contains a number of woodcuts of foreign plants.*]

1583

Caesalpinus, Andreas [Cesalpino, Andrea]. De plantis libri xvi. ... Florentiae, Apud Georgium Marescottum. 1583.

Clusius, Carolus [l'Écluse or l'Escluse, Charles de]. Car. Clusii atrebatis Rariorum aliquot Stirpium, per Pannoniam, Austriam, et vicinas ...Historia.... Antverpiae Ex officina Christophori Plantini. 1583.

Dodonaeus, Rembertus [Dodoens, Rembert]. Stirpium historiae pemptades sex. sive libri xxx. Antverpiae, Ex officina Christophori Plantini. 1583.

1584

Linocier, Geofroy. L'histoire des plantes, traduicte de latin en françois:...à Paris, Chez Charles Macé.... 1584.

1585

Durante, Castor. Herbario Nuovo.... Roma, Per Iacomo Bericchia, e Iacomo Turnierii. 1585. [*Another copy bearing same date, Appresso Bartholomeo Bonfadino et Tita Diani.*]

—— Hortulus sanitatis,...Ein...Gåhrtlin der Gesundtheit...Durch Petrum Uffenbachium,... Getruckt zu Frankfort am Måyn, durch Nicolaum Hoffmann. 1609.

1586

Matthiolus, Petrus Andreas [Mattioli, Pierandrea]. De plantis Epitome utilissima...aucta et locupletata, à D. Ioachimo Camerario,... accessit,...liber singularis de itinere...in Baldum montem...auctore Francisco Calceolario Francofurti ad Moenum. 1586.

—— Kreuterbuch...Jetzundt...gemehret und verfertiget Durch Ioachimum Camerarium.... Getruckt zu Franckfurt am Mayn. 1586.

280

Appendix I

Le Moyne de Morgues, Jacques. La Clef des Champs,... Imprimé aux Blackefriers. 1586.

1586, 7

Dalechampius, Jacobus [d'Aléchamps or Daléchamps, Jacques]. Historia generalis plantarum,... Lugduni, apud Gulielmum Rovillium. 1586, 7.

1588

Camerarius, Joachim. Hortus medicus et philosophicus:... Francofurti ad Moenum. 1588.

—— Icones accurate...delineatae praecipuarum stirpium, quarum descriptiones tam in Horto.... Impressum Francofurti ad Moenum. 1588. [*These figures are generally bound up with the* Hortus medicus.]

Porta, Johannes Baptista [Porta, Giambattista]. Phytognomonica.... Neapoli, Apud Horatium Saluianum. 1588.

1588, 91

Tabernaemontanus or Theodorus, Jacobus [Dietrich, Jacob]. Neuw Kreuterbuch,... Franckfurt am Mayn, 1588. (*Colophon:*) durch Nicolaum Bassaeum. 1591.

—— (*Another Edition*). Eicones plantarum seu stirpium. Nicolaus Bassaeus, Francofurti ad Moenum, 1590. [*This edition contains the figures only.*]

—— (*Another Edition*). Neuw vollkommentlich Kreuterbuch,... gemehret, Durch Casparum Bauhinum.... Franckfurt am Mayn, Durch Nicolaum Hoffman, In verlegung Johannis Bassaei und Johann Dreutels. 1613.

1592

Alpinus, Prosper [Alpino, Prospero]. De plantis Aegypti.... Venetiis...Apud Franciscum de Franciscis Senensem. 1592.

Columna, Fabius [Colonna, Fabio]. ΦΥΤΟΒΑΣΑΝΟC sive plantarum aliquot historia.... Ex Officina Horatii Saluiani. Neapoli, 1592. Apud Io. Iacobum Carlinum, et Antonium Pacem.

Zaluziansky von Zaluzian, Adam. Methodi herbariae, libri tres. Pragae, in officina Georgii Dacziceni. 1592.

1596

Bauhinus, Casparus [Bauhin, Gaspard]. ΦΥΤΟΠΙΝΑΞ seu enumeratio plantarum.... Basileae, per Sebastianum Henricpetri. 1596.

281

Appendix I

1597

Gerard, John [Gerarde, John]. The Herball or Generall Historie of Plantes.... Imprinted at London by John Norton. 1597.

—— (*Another Edition*). The Herball or Generall Historie of Plantes.... Very much Enlarged and Amended by Thomas Johnson Citizen and Apothecarye of London. London, Printed by Adam Islip, Joice Norton and Richard Whitakers. 1633.

—— (*Another Edition, without essential alteration*). 1636.

1601

Bauhinus, Casparus [Bauhin, Gaspard]. Animadversiones in historiam generalem plantarum Lugduni editam.... Francoforti, Excudebat Melchior Hartmann, Impensis Nicolai Bassaei.... 1601.

Clusius, Carolus [l'Écluse or l'Escluse, Charles de]. Caroli Clusii atrebatis,...rariorum plantarum historia.... Antverpiae Ex officina Plantiniana Apud Joannem Moretum. 1601.

1606

Columna, Fabius [Colonna, Fabio]. Minus cognitarum stirpium aliquot, ΕΚΦΡΑΣΙC.... Romae. Apud Guilielmum Facciottum. 1606. Pars altera. Romae. Apud Jacobum Mascardum. 1616.

Spigelius, Adrianus [Spieghel, A.]. Isagoges in rem herbariam Libri Duo.... Patavii, Apud Paulum Meiettum. Ex Typographia Laurentii Pasquati. 1606.

1611

Renealmus, Paulus [Reneaulme, Paul]. Specimen Historiae Plantarum. Parisiis, Apud Hadrianum Beys.... 1611.

1612

Bry, Johannes Theodorus de. Florilegium Novum. Cive Oppenheimense. 1612.

—— (*Another Edition*). Florilegium Renovatum. Prostat Francofurti apud Matthaeum Merianum, 1641.

1613

Beslerus, Basilius [Besler, Basil]. Hortus Eystettensis.... [Eichstadt.] 1613.

282

Appendix I

1614

Passaeus, Crispianus [Passe, C. de, or Pas, Crispijn vande].
Hortus floridus...Extant Arnhemii. Apud Ioannem Ianssonium... 1614.

—— (*Another Edition*). A Garden of Flowers.... Printed at Utrecht,
By Salomon de Roy. 1615.

1615

Hernandez, Francisco. Quatro libros. de la naturaleza, y virtudes
de las plantas,...en el uso de Medicina en la Nueva España...Traduzido
...Francisco Ximenez.... En Mexico...Viuda de Diego Lopez Davalos.
...1615.

1616

Olorinus, Johannes [Sommer, Johann, aus Zwickau]. Centuria
Herbarum Mirabilium Das ist: Hundert Wunderkråuter.... Magdeburgk, Bey Levin Braunss.... 1616.

—— Centuria Arborum Mirabilium Das ist: Hundert Wunderbäume.
... Magdeburgk, Bey Levin Braunss.... 1616.

1619

Bauhinus, Joannes [Bauhin, Jean] and **Cherlerus, J. H.** [Cherler,
J. H.]. J. B....et J. H. C....historiae plantarum generalis...prodro-
mus.... Ebroduni, Ex Typographia Societatis Caldorianae. 1619.

(*A more complete version*). Historia plantarum universalis...Quam
recensuit et auxit...Chabraeus...publici fecit, Fr. Lud. a Graffenried....
Ebroduni, 1650,1.

1620

Bauhinus, Casparus [Bauhin, Gaspard]. ΠΡΟΔΡΟΜΟΣ Theatri
botanici.... Francofurti ad Moenum, Typis Pauli Jacobi, impensis Ioannis
Treudelii. 1620.

1622

Anon. (Rabel, Daniel). Theatrum Florae. Apud Nicolaum Matho-
niere. 1622.[1]

—— (*Another Edition*). Theatrum Florae In quo Ex toto Orbe selecti
Mirabiles...Lutetiae Parisiorum Apud Petrum Firens. 1633.

1623

Bauhinus, Casparus [Bauhin, Gaspard]. ΠΙΝΑΞ theatri botanici....
Basileae Helvet. Sumptibus et typis Ludovici Regis. 1623.

[1] The writer has not seen a copy of this edition, nor of the edition of 1627; on this
work see Savage (1923[1]), Appendix II.

Appendix I

1825

Popp, Johann [Poppe, Johann]. Kråuter Buch...nach rechter art der Signaturen der himlischen Einfliessung nicht allein beschrieben,... Leipzig, In Verlegung Zachariae Schůrers, und Matthiae Gőtzen.... 1625.

1628

Brosse, Guy de la. De la nature, vertu, et utilité des plantes.... A Paris, Chez Rollin Baragnes.... 1628.

1629

Johnson, Thomas. Iter plantarum investigationis Ergô Susceptum *A decem Sociis, in Agrum* Cantianum. Anno Dom. 1629. Iulii 13. Ericetum Hamstedianum Sive Plantarum ibi crescentium observatio habita, Anno eodem I. *Augusti* Descripta studio, et operâ *Thomae Iohnsoni.* [*Only copy known to the writer is at Magdalen College, Oxford.*]

—— (*Another Edition*). Descriptio Itineris Plantarum...in Agrum Cantianum...et Enumeratio Plantarum in Ericeto Hampstediano locisque vicinis Crescentium.... Excudebat, Tho. Cotes. [London.] 1632.

Parkinson, John. Paradisi in Sole Paradisus Terrestris. A Garden of all sorts of pleasant flowers which our English ayre will permitt to be noursed up;... (*Colophon:*) London, Printed by Humfrey Lownes and Robert Young at the signe of the Starre on Bread-street hill. 1629.

1631

Donati, Antonio. Trattato de semplici,... in Venetia,...Appresso Pietro Maria Bertano. 1631.

1634

Johnson, Thomas. Mercurius Botanicus:... Londini, Excudebat Thom. Cotes. 1634.

1640

Parkinson, John. Theatrum botanicum: The Theater of Plants. Or, an Herball of a Large Extent.... London, Printed by Tho. Cotes. 1640. [*The wording of the engraved half-title differs from the title-page.*]

1649

Culpeper, Nicholas. A Physicall Directory or A translation of the London Dispensatory Made by the Colledge of Physicians in London... with many hundred additions.... London, Printed for Peter Cole.... 1649.

—— (*Another Edition*). The English Physitian enlarged.... London, Printed by Peter Cole.... 1653.

Appendix I

1650

[How, William.] Phytologia Britannica, natales exhibens Indigenarum stirpium sponte Emergentium. Londoni, Typis Ric. Cotes, Impensis Octaviani Pulleyn. 1650.

1656

Cole, William [Coles, William]. The Art of Simpling. London, Printed by J. G. for Nath: Brook. 1656.

1657

Cole, William [Coles, William]. Adam in Eden: or, Natures Paradise.... London, Printed by J. Streater, for Nathaniel Brooke.... 1657.

1658

Bauhinus, Casparus [Bauhin, Gaspard]. Caspari Bauhini...Theatri botanici sive historiae plantarum...liber primus editus opera et cura Io. Casp. Bauhini. Basileae. Apud Joannem König. 1658.

1659

Lovell, Robert. ΠΑΜΒΟΤΑΝΟΛΟΓΙΑ, sive Enchiridion botanicum, or a compleat Herball.... Oxford, Printed by William Hall, for Ric. Davis.... 1659.

1662

Jonstonus, Johannes [Jonston or Johnstone, John]. Dendrographias sive historiae naturalis de arboribus et fruticibus...libri decem.... Francofurti ad Moenum. Typis Hieronymi Polichii. Sumptibus Haeredum Matthaei Meriani. 1662.

1664

Turner, Robert. ΒΟΤΑΝΟΛΟΓΙΑ. The Brittish Physician: or, The Nature and Vertues of English Plants. London, Printed by R. Wood for Nath. Brook. 1664.

1666

Chabraeus, Dominicus [Chabrey, D.]. Stirpium icones et sciagraphia....Genevae, Typis Phil. Gamoneti et Iac. de la Pierre. 1666.

1667, 8

Aldrovandus, Ulysses [Aldrovandi, Ulisse]. Ulyssis Aldrovandi... Dendrologiae naturalis scilicet arborum historiae libri duo.... Bononiae typis Io. Baptistae Ferronii. 1667, 8. [It is doubtful whether this posthumous book, which was edited by O. Montalbani, should appear under Aldrovandi's name.]

1670

Nylandt, Petrus. De Nederlandtse Herbarius of Kruydt-Boeck,... t'Amsterdam, voor Marcus Doornick,... 1670.

APPENDIX II

AN ALPHABETICAL LIST OF THE HISTORICAL AND CRITICAL
WORKS CONSULTED DURING THE PREPARATION OF THIS
BOOK

Adam, M. (1620) Vitae Germanorum Medicorum. Heidelberg.

Adanson, M. (1763) Familles des Plantes. Contenant une Préface Istorike sur l'état ancien et actuel de la Botanike. Paris.

Alcock, R. H. (1876) Botanical Names for English Readers. London.

Arber, A. (1913[1]) The Botanical Philosophy of Guy de la Brosse. Isis, vol. I, pp. 361–9.

Arber, A. (1913[2]) Nehemiah Grew, 1641–1712. In Makers of British Botany, edited by F. W. Oliver. Cambridge.

Arber, A. (1921) The Draughtsman of the Herbarum Vivae Eicones. Journ. Bot., vol. LIX, pp. 131–2.

Arber, A. (1928) On a French version of the herbal of Leonard Fuchs. Notes and Queries, vol. CLIV, pp. 381–3.

Arber, A. (1931) Edmund Spenser and Lyte's "Nievve Herball". Notes and Queries, vol. CLX, pp. 345–7.

Arber, A. (1936) A Recent Discovery in Sixteenth Century Botany. Nature, vol. CXXXVII, pp. 258–9.

Avoine, P. J. d' See **Morren, C.**

Baldensperger, L. See **Crowfoot, G. M.**

Banchi, L. See **Fabiani, G.**

Barlow, H. M. (1913) Old English Herbals. 1525–1640. Proc. Roy. Soc. Med., vol. VI (Sect. Hist. Med.), pp. 108–49.

Bauhin, J. (1591) De Plantis à Divis Sanctis....Additae sunt Conradi Gesneri Medici Clariss. Epistolae...à Casparo Bauhino. Basileae Apud Conrad. Waldkirch.

Béguinot, A. (1909) Flora Padovana, pt. I. Padua.

286

Appendix II

Bellini, R. (1898)   Gli autografi dell' "Ekphrasis" di Fabio Colonna. Nuovo Giornale Bot. Ital., vol. v, N.S., pp. 45–56.

Berendes, J. (1902)   Des Pedanios Dioskurides aus Anazarbos Arzneimittellehre...übersetzt...von. J. Berendes. Stuttgart.

Britten, J. (1881)   The Names of Herbes. by William Turner. A.D. 1548. Edited by James Britten. London.

Britten, J. &   A Dictionary of English Plant-names. English
Holland, R. (1886) Dialect Society. London.

Broadwood, L. E.   The Magical Herb Wormwood in Switzerland.
(1925)   Folklore, vol. XXXVI, pp. 387–8.

Brosig, M. (1883)   Die Botanik des älteren Plinius. Kgl. evangel. Gymnasium zu Graudenz. xvii. Jahresber. über das Schuljahr Ostern 1882 bis Ostern 1883. Programm 32., pp. 1–30.

Camus, G. (1886)   L' Opera Salernitana "Circa Instans" ed il testo primitivo del "Grant Herbier en Francoys". Memorie della Regia Accademia di Scienze, Lettere ed Arti in Modena, ser. II, vol. IV, Mem. della Sezione di Lettere, pp. 49–199.

Camus, J. (1894)   Les Noms des Plantes du Livre d'Heures d'Anne de Bretagne. Journ. de Bot., vol. VIII, pp. 325–35, 345–52, 366–75, 396–401.

Camus, J. (1895)   Historique des premiers herbiers. Malpighia, Anno 9, pp. 283–314.

Candolle, C. de   L'Herbier de Gaspard Bauhin déterminé par
(1904)   A. P. de Candolle. Bull. de l'herbier Boissier, ser. II, vol. IV, pp. 201–16, 297–312, 458–74, 721–54.

Choate, H. A. (1917) The Earliest Glossary of Botanical Terms; Fuchs 1542. Torreya, vol. XVII, pp. 186–201.

Choulant, L. (1832)   Macer Floridus de viribus herbarum...secundum codices manuscriptos...recensuit...Ludovicus Choulant.... Lipsiae.

Choulant, L.   Handbuch der Bücherkunde für die aeltere
(1841, 1926)   Medizin. 2nd ed. Leipsic, 1841. (Reprinted in facsimile, 1926.)

Appendix II

Choulant, L.
(1857, 1858, 1924) Botanische und anatomische Abbildungen des Mittelalters. Archiv für die zeichnenden Künste. Jahrg. III, pp. 188–309. (Reprinted in 1858 as Graphische Incunabeln für Naturgeschichte und Medizin; and in facsimile under this title in 1924, Leipsic.)

Christ, H. (1912) Die illustrierte spanische Flora des Carl Clusius vom Jahre 1576. Österreich. bot. Zeitschrift, vol. LXII, pp. 132–5, 189–94, 229–38, 271–5.

Christ, H.
(1912, 13) Die ungarische-österreichische Flora des Carl Clusius vom Jahre 1583. Österreich. bot. Zeitschrift, vol. LXII, pp. 330–4, 393–4, 426–30; vol. LXIII, pp. 131–6, 159–67.

Christ, H. (1913) Eine Basler Flora von 1622. Basler Zeitschrift für Geschichte und Alterthumskunde, vol. XII, pp. 1–15.

Christ, H. (1927) Otto Brunfels und seine Herbarum vivae eicones. Ein botanischer Reformator des XVI. Jahrhunderts. Verhandl. d. Naturforsch. Gesellsch. Basel, vol. XXXVIII, pp. 1–11.

Church, A. H.
(1919) Brunfels and Fuchs. Journ. Bot., vol. LVII, pp. 233–44.

Clarke, W. A.
(1900) First Records of British Flowering Plants. Edition 2. London.

Cockayne, T. O.
(1864) Leechdoms, Wortcunning, and Starcraft of Early England. Chronicles and Memorials of Great Britain and Ireland during the Middle Ages. Rolls Series, vol. I.

Copinger, W. A.
(1895, 1898, 1902) Supplement to Hain's Repertorium Bibliographicum. London.

Courtois, R. (1835) Commentarius in Remberti Dodonaei Pemptades. Nova Acta Physico-Medica Acad. Caes. Leopold.-Carol. Naturae Curiosorum, vol. XVII, pt. II, pp. 763–840.

Crane, W. (1906) Of the Decorative Illustration of Books Old and New. London.

Crowfoot, G. M. &
Baldensperger, L. From Cedar to Hyssop: A Study in the Folklore of Plants in Palestine. London.
(1932)

Appendix II

Cuvier, G. (1841)   Histoire des Sciences Naturelles, depuis leur origine jusqu'à nos jours, complétée...par M. Magdeleine de Saint Agy, pt. ii, vol. ii. Paris.

Daubeny, C. (1857)   Lectures on Roman Husbandry. Oxford.

Degeorge, L. (1886)   La Maison Plantin à Anvers. Édition 3. Bruxelles.

Denucé, J.   See **Rooses, M.**

Dorveaux, P. (1913)   Le Livre des Simples Medecines. Traduction française du *Liber de simplici medicina dictus Circa instans* de Platearius tirée d'un manuscrit du xiiie siècle. Publications de la Soc. franç. d'histoire de la médecine, No. 1. Paris.

Downes, H. (1917)   Henry Lyte of Lyte's Cary. Notes and Queries for Somerset and Dorset, vol. xv, pp. 157–9.

Drewitt, F. D. (1928)   The Romance of the Apothecaries' Garden at Chelsea. 3rd edition. Cambridge.

Druce, G. C. (1923)   Herbaria. Rept. Bot. Exchange Club, vol. vi, pt. 5 (1923 for 1922), pp. 756–66.

Emmanuel, E. (1912)   Étude comparative sur les plantes dessinées dans le *Codex Constantinopolitanus* de *Dioscoride*. Travail exécuté dans l'Institut pharmaceutique de l'Université de Berne et l'Herbier Boissier à Chambésy près Genève. Schweiz. Wochenschrift f. Chemie und Pharmazie: Journal suisse de Chimie et Pharmacie, Zurich, vol. lxi (Jahrg. 50), pp. 45–50, 64–72.

Emmart, E. W. (1935)   Concerning the Badianus Manuscript, an Aztec Herbal, "Codex Barberini, Latin 241" (Vatican Library). Smithsonian Misc. Collections, vol. xciv, No. 2, 14 pp.

Fabiani, G. (1872)   La Vita di Pietro Andrea Mattioli, edited by L. Banchi. Siena.

Faraglia, N. (1885)   Fabio Colonna Linceo. Archivio Storico per le Province Napoletane. Anno 10, pp. 665–749.

Fellner, S. (1881)   Albertus Magnus als Botaniker. Jahres-Ber. des kais. kön. Ober-Gymnasiums zu den Schotten in Wien, pp. 1–90.

Appendix II

Field, H. (1878)   Memoirs of the Botanic Garden at Chelsea belonging to the Society of Apothecaries of London. Revised by R. H. Semple. London.

Fischer, H. (1929)   Mittelalterliche Pflanzenkunde. Munich.

Forster, E. S. (1927) The Turkish Letters of Ogier Ghiselin de Busbecq. Oxford.

Gabrieli, G. (1929)   Due codici iconografici di piante miniate nella Biblioteca Reale di Windsor. Rendiconti della r. Accad. Naz. dei Lincei. Rome. Classe di Sci. fis. mat. e nat., vol. x, ser. 6a, 2nd sem., fasc. 10, pp. 531–8.

Gaselee, S. (1925)   Joyfull Newes out of the Newe Founde Worlde written in Spanish by Nicolas Monardes physician of Seville and Englished by John Frampton, Merchant anno 1577. With an Introduction by S. Gaselee. London.

George, W. ([1880]) Lytes Cary Manor House, Somerset. Bristol. n.d.

Gesamtkatalog der Wiegendrucke   See **Kommission f.d.G.d.W.** (1925 onwards).

Giacosa, P. (1901)   Magistri Salernitani nondum editi. Catalogo ragionato della esposizione di storia della medicina aperta in Torino nel 1898. Torino. [In 2 parts, text and atlas.]

Gibson, S. (1931–2)   Fragments from Bindings at Queen's College, Oxford. Trans. Bibl. Soc. (The Library), ser. iv, vol. xii, pp. 429–33.

Greene, E. L. (1909)   Landmarks of Botanical History. A Study of Certain Epochs in the Development of the Science of Botany. Pt. i. Prior to 1562 A.D. Smithsonian Misc. Coll. No. 1870. Pt. of vol. LIV. Washington.

Gunther, R. T. (1922)   Early British Botanists and their Gardens based on unpublished writings of Goodyer, Tradescant, and others. Oxford.

Appendix II

Gunther, R. T. (1925)	The Herbal of Apuleius Barbarus from the early twelfth-century manuscript formerly in the Abbey of Bury St Edmunds (MS. Bodley 130) described by R. T. Gunther. Oxford, for the Roxburghe Club.
Gunther, R. T. (1934)	The Greek Herbal of Dioscorides illustrated by a Byzantine A.D. 512, Englished by John Goodyer, A.D. 1655, Edited and printed A.D. 1933 (*sic*) by R. T. Gunther. Oxford.
Hain, L. (1826, 27, 31, 38)	Repertorium Bibliographicum...ad annum MD. Stuttgart, Tübingen and Paris.
Haller, A. von (1771–2)	Bibliotheca botanica. Tiguri.
Hanhart, J. (1824)	Conrad Gessner. Ein Beytrag zur Geschichte... im 16ten Jahrhundert. Winterthur.
Hartmann, F. (1896)	The Life of Philippus Theophrastus Bombast of Hohenheim known by the name of Paracelsus. 2nd edition. London.
Hatton, R. G. (1909)	The Craftsman's Plant-Book. London.
Henslow, G. (1899)	Medical Works of the Fourteenth Century together with a List of Plants Recorded in Contemporary Writings, with their Identifications. London.
Heron-Allen, E. (1928)	Barnacles in Nature and in Myth. Oxford.
Hess, J. W. (1860)	Kaspar Bauhin's...Leben und Character. Basel.
Hett, W. S. (1936)	Aristotle. Minor Works. Translated by W. S. Hett. London.
Hill, A. W. (1915)	The History...of Botanic Gardens. Ann. Missouri Bot. Gard., vol. II, pp. 185–240.
Hill, A. W. (1937)	Preface by Sir Arthur Hill to Turrill, W. B., A Contribution to the Botany of Athos Peninsula Kew Bull., pp. 197–8.
Hizlerus, G. (1566)	Oratio de vita et morte Leonharti Fuchsii. Tubingae.
Holland, P. (1601)	The Historie of the World. Commonly called, the naturall historie of C. Plinius Secundus. Translated by Philemon Holland. London.

Appendix II

Holland, R.	See **Britten, J.**
Hort, A. (1916)	Theophrastus. Enquiry into Plants. With an English Translation by Sir Arthur Hort. London.
Howald, E. & **Sigerist, H. E.** (1927)	Antonii Musae de Herba Vettonica Liber. Pseudoapulei Herbarius, etc. Leipsic.
Hunger, F. W. T. (1917¹)	Catalogus van de Tentoonstelling gehouden te Leiden 29 Juni 1917, ter Gelegenheid van den 400sten Geboortedag van Rembertus Dodonaeus. Leiden.
Hunger, F. W. T. (1917²)	Dodonée comme botaniste. Janus, année 22, pp. 153–62.
Hunger, F. W. T. (1927)	Charles de l'Escluse (Carolus Clusius), 1526–1609. Janus, année 31, pp. 139–51.
Hunger, F.W. T. (1935¹)	Catalogue of Botanical Incunabula and Post-incunabula. Exhibition of Books. VIth Intern. Bot. Congress. Amsterdam, Sept. 2–7, 1935.
Hunger, F. W. T. (1935²)	The Herbal of Pseudo-Apuleius from the ninth-century Manuscript in the Abbey of Monte Cassino [Codex Casinensis 97] together with the first printed edition of Joh. Phil. de Lignamine [Editio Princeps Romae 1481]. Leyden.
Irmisch, T. (1859)	Literaturgeschichtl. Bemerkung über eine Ausgabe von dem Kräuterbuche des Tragus. Bot. Zeit., Jahrg. 17, pp. 30–1.
Irmisch, T. (1862)	Ueber einige Botaniker des 16. Jahrhunderts. Öff. Prüfung des F. Schwartzburg. Gymnasiums zu Sondershausen, pp. 3–58.
Istvánffi, G. de (1898–1900)	Caroli Clusii Atrebatis Icones Fungorum in Pannonis Observatorum sive Codex Clusii Lugduno Batavensis...cura et sumptibus Dr¹ˢ Gy. de Istvánffi. Études et Commentaires sur le Code de l'Escluse augmentés de quelques notices Biographiques. Budapest.
Jackson, B. D. (1876)	A Catalogue of Plants cultivated in the Garden of John Gerard, In the years 1596–1599. Edited with...a life of the author, by B. D. Jackson. London.

292

Appendix II

Jackson, B. D.
(1877)
Libellus de re herbaria novus, by William Turner, originally published in 1538, reprinted in facsimile, with notes, modern names, and a life of the author by B. D. J. London.

Jackson, B. D.
(1881)
Guide to the Literature of Botany. London.

Jackson, B. D.
(1906)
The History of Botanic Illustration. Trans. Hertfordshire Nat. Hist. Soc., vol. xii, pp. 145–56, 1906, for 1903–5.

Jackson, B. D.
(1924)
Botanical Illustration from the Invention of Printing to the Present Day. Journ. Roy. Hort. Soc., vol. xlix, pp. 167–77.

Jardine, W. (1843)
The Naturalist's Library. Edited by Sir W. Jardine, vol. xii, Memoir of Gesner. Edinburgh.

Jayne, K. G. (1910)
Vasco da Gama and his successors, 1460–1580. London.

Jessen, K. F. W.
(1864)
Botanik der Gegenwart und Vorzeit in cultur-historischer Entwickelung. Leipzig.

Jorge, R. (1916)
Comentos...Amato Lusitano. Pòrto.

Kaestner, H. F.
(1896)
Pseudo-Dioscoridis de herbis femininis. Hermes, vol. xxxi, pp. 578–636.

Karabacek, J. de
(1906)
Dioskurides. Codex Aniciae Julianae picturis illustratus, nunc Vindobonensis. Med. Gr. I. phototypice editus. Lugduni Batavorum.

Kessler, H. F. (1870)
Das älteste und erste Herbarium Deutschlands, im Jahr 1592 von Dr Caspar Ratzenberger angelegt...beschrieben...von Dr H. F. Kessler. Cassel.

Kew, H. W. &
Powell, H. E. (1932)
Thomas Johnson, Botanist and Royalist. London.

Kickx, J. (1838)
Esquisses sur les ouvrages de quelques anciens naturalistes belges. I. Auger-Gislain Busbecq. Bull. de l'acad. roy. des sciences et belles-lettres de Bruxelles, vol. v, pp. 202–15.

Killermann, S.
(1909)
Zur ersten Einführung amerikanischer Pflanzen im 16. Jahrhundert. Naturwissenschaftliche Wochenschrift, vol. xxiv (N.S., vol. viii), pp. 193–200.

Appendix II

Killermann, S.   A. Dürer's Pflanzen- und Tierzeichnungen.
(1910)   Studien zur deutschen Kunstgeschichte. Heft 119. Strasburg.

Killermann, S.   Albrecht Dürer als Naturfreund. Der Aar
(1911)   (Regensburg), Jahrg. 1, Heft 6, pp. 751–65.

Klebs, A. C.   Herbals of the Fifteenth Century. Incunabula
(1917, 18)   Lists I. Papers of the Bibliographical Society of America, vol. xi, Nos. 3–4, pp. 75–92; vol. xii, Nos. 1–2, pp. 41–57.

Klebs, A. C. (1925)   Herbal Facts and Thoughts. Reprint of an introduction to the Catalogue of Early Herbals from the Library of Dr Karl Becher. L'Art Ancien S.A. Lugano.

Klebs, A. C. (1926)   (Letter) Hortus Sanitatis. Times Lit. Sup., Aug. 5.

Klebs, A. C. (1932)   Gleanings from Incunabula of Science and Medicine. Papers of the Bibliographical Society of America, vol. xxvi, pp. 52–88.

Kommission f.d.G.   Gesamtkatalog der Wiegendrucke. (In progress.)
d.W. (1925 onwards)   Leipzig.

Kristeller, P. (1905)   Kupferstich und Holzschnitt in vier Jahrhunderten. Berlin.

Langkavel, B. (1866) Botanik der spaeteren Griechen. Berlin.

Legré, L.   La Botanique en Provence au xvie siècle: Pierre
(1899–1904)   Pena et Mathias de Lobel, 1899. Félix et Thomas Platter. 1900[1]. Leonard Rauwolff. Jacques Raynaudet. 1900[2]. Louis Anguillara. Pierre Belon. Charles de l'Escluse. Antoine Constantin. 1901. Les deux Bauhin, Jean-Henri Cherler et Valerand Dourez. 1904. Marseille.

Locy, W. A. (1921)   The Earliest Printed Illustrations of Natural History. Sci. Monthly, New York, vol. xiii, pp. 238–58.

Lones, T. E. (1912)   Aristotle's Researches in Natural Science. London.

Macfarlane, J.   Antoine Vérard. Bibl. Soc. Illustrated Mono-
(1900)   graphs, No. 7, 1900 for 1899. London.

Magdeleine de Saint Agy, T. See **Cuvier, G.**

Appendix II

Maiwald, V. (1904) Geschichte der Botanik in Böhmen. Wien und Leipzig.

Markham, C. (1913) Colloquies on the Simples and Drugs of India by Garcia da Orta (New edition, Lisbon, 1895, edited by the Conde de Ficalho) translated by Sir Clements Markham. London.

Mattirolo, O. (1897) L' Opera Botanica di Ulisse Aldrovandi (1549–1605). Bologna.

Mattirolo, O. (1898) La Nuova *Sala Aldrovandi* nell' Istituto botanico della R. Università di Bologna. Malpighia, Anno 12, pp. 140–54.

Mattirolo, O. (1899) Illustrazione del primo Volume dell' Erbario di Ulisse Aldrovandi. Genova.

Mattirolo, O. (1907) Parole...in occasione delle onoranze per Ulisse Aldrovandi nel III centenario dalla sua morte. Atti d. Accad. R. delle Sci. di Torino (Anno 1906–7), vol. XLII, pp. 1037–40.

Maxwell-Lyte, H. C. (1892) The Lytes of Lytescary. Proc. Somerset. Arch. and Nat. Hist. Soc. vol. XXXVIII, pt. II, pp. 1–110.

Meerbeck, P. J. van (1841) Recherches historiques et critiques sur la vie et les ouvrages de Rembert Dodoens (Dodonaeus). Malines.

Meusnier de Querlon, A. G. (1774) Journal du voyage de Michel de Montaigne en Italie...en 1580 et 1581. Rome and Paris.

Meyer, E. H. F. (1854–7) Geschichte der Botanik. Königsberg.

Meyer, E. H. F. & Jessen, K. F. W. (1867) Alberti Magni ex ordine praedicatorum de vegetabilibus libri VII,...editionem criticam ab Ernesto Meyero coeptam absolvit Carolus Jessen. Berlin.

Miall, L. C. (1912) The Early Naturalists. Their Lives and Work (1530–1789). London.

Milt, B. (1936) Notizen zur schweizerischen Kulturgeschichte von H. Schinz und K. Ulrich. 102. Conrad Gessner's *Historia Plantarum* (Fragmenta relicta). Vierteljahrsschrift der naturf. Gesellsch. in Zürich, Jahrg. 81, pp. 285–91.

Moehsen, J. C. W. (1783) Beiträge zur Geschichte der Wissenschaften in der Mark Brandenburg von den ältesten Zeiten an bis zu Ende des sechszehnten Jahrhunderts. I. Leben Leonhard Thurneissers zum Thurn. Berlin und Leipzig.

Mohl, H. von (1859) Einige Worte zur Rechtfertigung von Leonhard Fuchs. Bot. Zeit., Jahrg. 17, p. 189.

Montaigne, M. de See **Meusnier de Querlon, A. G.**

Morren, C. (1851) Prologue consacré à la mémoire de Rembert Dodoëns. La Belgique Horticole, Liége, vol. i.

Morren, C. (1853) Prologue consacré à la mémoire de Charles de l'Escluse. La Belgique Horticole, Liége, vol. iii.

Morren, C. & d'Avoine, P. J. (1850) Éloge de Rembert Dodoëns,...suivi de la Concordance des espèces végétales décrites et figurées par Rembert Dodoëns avec les noms que Linné et les auteurs modernes leur ont donnés. Malines.

Morren, É. (1875¹) Charles de l'Escluse, sa vie et ses œuvres. 1526–1609. Bull. de la Féd. des Soc. d'Hort. de Belgique. Liége (1874). (There are some original notes in a review of this work by B. D. Jackson, Journ. Bot., vol. xiii (New Ser. vol. iv), 1875, pp. 347–9.)

Morren, É. (1875²) Mathias de l'Obel, sa vie et ses œuvres, 1538–1616. Bull. de la Féd. des Soc. d'Hort. de Belgique. Liége.

Muther, R. (1884) Die deutsche Bücherillustration der Gothik und Frührenaissance (1460–1530). München und Leipzig.

Nelmes, E. See **Sprague, T. A.**

Netter, W. See **Peters, H.**

Olmedilla y Puig, J. (? 1896) El sabio médico Portugués del siglo XVI García da Orta. [? Madrid.]

Olmedilla y Puig, J. (1897) Estudio histórico...Nicolás Monardes. Madrid.

Olmedilla y Puig, J. (1899) Estudio histórico...Cristóbal Acosta. Madrid.

Appendix II

Parkinson, J. (1904) Paradisi in Sole Paradisus Terrestris. Faithfully reprinted from the edition of 1629. London.

Payne, J. F. (1885) Old Herbals: German and Italian. The Mag. of Art, vol. VIII, pp. 362–8.

Payne, J. F. (1903) On the "Herbarius" and "Hortus Sanitatis". Trans. Bibliographical Soc., vol. VI, 1903 (for 1900–2), pp. 63–126.

Payne, J. F. (1908) English Herbals (Summary of a paper). Trans. Bibl. Soc., vol. IX, 1908 (for 1906–8), pp. 120–3.

Payne, J. F. (1912) English Herbals. Trans. Bibliographical Soc., vol. XI, 1912 (for 1909–11), pp. 299–310. (This article is a posthumous reprint of Payne, J. F. (1908) with figures.)

Penzig, O. (1905) Contribuzioni alla Storia della Botanica. I. Illustrazione degli Erbarii di Gherardo Cibo. II. Sopra un Codice miniato della Materia Medica di Dioscoride. Milan.

Petermann, W. L. See **Richter, H. E.**

Peters, H. (1889) Pictorial History of Ancient Pharmacy....Trans. by Dr W. Netter. Chicago.

Pfeiffer, F. (1861) Das Buch der Natur von Konrad von Megenberg...herausgegeben von Dr Franz Pfeiffer. Stuttgart.

Pitton de Tournefort, J. (1694) Elemens de Botanique, vol. I. Paris.

Pitton de Tournefort, J. (1700) Institutiones rei herbariae. Ed. 2, vol. I. Paris.

Planchon, J. E. & G. (1866) Rondelet et ses Disciples ou la Botanique à Montpellier au XVIme Siècle. Appendix by J. E. and G. Planchon. Montpellier.

Pouchet, F. A. (1853) Histoire des sciences naturelles au moyen âge ou Albert le Grand et son époque, considérés comme point de départ de l'école expérimentale. Paris.

Powell, H. E. See **Kew, H. W.**

Prideaux, W. R. B. (1926) (Letters) Hortus Sanitatis. Times Lit. Sup., July 22 and August 19.

Appendix II

Prior, R. C. A.
(1879)
On the Popular Names of British Plants. 3rd edition. London.

Pritzel, G. A. (1846) Meister Johann Wonnecke von Caub. Bot. Zeit., Jahrg. 4, pp. 785–90.

Pritzel, G. A.
(1872, 7)
Thesaurus literaturae botanicae...editionem novam reformatam....Lipsiae.

Pryme, A. de la
(1870)
The Diary of Abraham de la Pryme, the Yorkshire Antiquary. Surtees Society's Publications, vol. LIV (for 1869). Durham, London and Edinburgh.

Pulteney, R. (1790) Historical and Biographical Sketches of the Progress of Botany in England, from its Origin to the Introduction of the Linnaean System. London.

Ralph, T. S. (1847) Opuscula omnia botanica Thomae Johnsoni nuperrime edita. London.

**Richter, H. E. &
Petermann, W. L.**
(1835, 40)
Caroli Linnaei Systema, Genera, Species plantarum,...seu Codex Botanicus Linneanus; ed. H. E. Richter; Lipsiae, 1835; Index alphabeticus, W. L. Petermann, Lipsiae, 1840.

Rooses, M.
(1882, 3)
Christophe Plantin imprimeur anversois. Antwerp. (Another edition in smaller format, 1896.)

Rooses, M. (1909) Catalogue of the Plantin-Moretus Museum. 2nd English edition. Antwerp.

**Rooses, M. &
Denucé, J.**
(1883–1918)
Correspondance de Christophe Plantin. Edited by M. Rooses, and (later) J. Denucé. Maatschappij der Antwerpische Bibliophilen. Antwerp.

Roth, F. W. E.
(1898)
Hieronymus Bock genannt Tragus (1498–1554). Bot. Centralbl. Jahrg. 19, vol. LXXIV, pp. 265–71, 313–18, 344–7.

Roth, F. W. E.
(1899[1])
Leonhard Fuchs, ein deutscher Botaniker, 1501–1566. Beihefte zum Bot. Centralbl. vol. VIII, pp. 161–91.

Roth, F. W. E.
(1899[2])
Jacob Theodor aus Bergzabern, genannt Tabernaemontanus. Ein deutscher Botaniker. Bot. Zeit., Jahrg. 57, pp. 105–23.

Roth, F. W. E.
(1900)
Otto Brunfels 1489–1534. Ein deutscher Botaniker. Bot. Zeit., Jahrg. 58, pp. 191–232.

Appendix II

Roth, F. W. E.
(1902)

Die Botaniker Eucharius Rösslin, Theodor Dorsten und Adam Lonicer 1526–1586. Centralbl. f. Bibliothekswesen, Jahrg. 19, pp. 271–86, 338–45.

Roze, E. (1899)

Charles de l'Escluse d'Arras le propagateur de la pomme de terre au xvie siècle. Sa biographie et sa correspondance. Paris.

Rytz, W. (1933)

Das Herbarium Felix Platters. Ein Beitrag zur Geschichte der Botanik des xvi. Jahrhunderts. Verhandl. d. Naturforsch. Gesellsch. in Basel, vol. xliv, pp. 1–222.

Rytz, W. (1936)

Pflanzenaquarelle des Hans Weiditz aus dem Jahre 1529: die Originale zu den Holzschnitten im Brunfels'schen Kräuterbuch. Bern.

Saccardo, P. A.
(1893)

Il primato degli Italiani nella Botanica. Malpighia, vol. vii, pp. 483–7.

Sachs, J. von (1890) History of Botany (1530–1860). Trans. by H. E. F. Garnsey, revised by I. B. Balfour. Oxford.

Saint-Lager, J. B.
(1885[1])

Histoire des Herbiers. Ann. Soc. Bot. de Lyon, année 13, pp. 1–120.

Saint-Lager, J. B.
(1885[2])

Recherches sur les Anciens Herbaria. Ann. Soc. Bot. de Lyon, année 13, pp. 237–81.

Salaman, R. N.
(1937)

The Potato in its Early Home. Journ. Roy. Hort. Soc. vol. lxii, pp. 61–77, 112–23, 153–62, 253–66.

Sarton, G.
(1927 onwards)

Introduction to the History of Science. Carnegie Institution, Washington. (In progress.)

Savage, S. (1921)

A Little-known Bohemian Herbal. Trans. Bibl. Soc. (The Library), ser. ii, vol. ii, pp. 117–31.

Savage, S. (1922)

The Discovery of some of Jacques le Moyne's Botanical Drawings. Gard. Chron., ser. iii, vol. lxxi, p. 44.

Savage, S. (1923[1])

The *Hortus Floridus* of Crispijn vande Pas the Younger. Trans. Bibl. Soc. (The Library), ser. ii, vol. iv, pp. 181–206.

Savage, S. (1923[2])

Early Botanic Painters. 1–6. Gard. Chron., ser. iii, vol. lxxiii, pp. 8, 92–3, 148–9, 200–1, 260–1, 336–7.

Savage, S.
(1928, 9)

Crispian Passaeus. Hortus Floridus. Text translated from the Latin by Spencer Savage. London.

Savage, S. (1935)

Studies in Linnaean Synonymy. I. Caspar Bauhin's "Pinax" and Burser's Herbarium. Proc. Linn. Soc. Lond., 148th session, 1935–6, pt. 1, 1935, pp. 16–26.

Savage, S. (1937)

Catalogue of the Manuscripts in the Library of the Linnean Society of London. Part II. Caroli Linnaei determinationes in hortum siccum Joachimi Burseri. London.

Schelenz, H. (1904)

Geschichte der Pharmazie. Berlin.

Schenck, H. (1920)

Martin Schongauers Drachenbaum. Naturwissenschaftliche Wochenschrift, vol. xxxv (N.S. vol. xix), pp. 775–80.

Schmid, A. (1936)

Zwei seltene Kräuterbücher aus dem vierten Dezennium des sechzehnten Jahrhunderts, nebst einem bibliographischen Verzeichnis aller bekannten frühen Ausgaben der Brunfelsschen Kräuterbücher als Anhang. Separat. aus dem Schweizerischen Gutenbergmuseum, No. 3. Bern.

Schmiedel, C. C.
(1751–4)

Conradi Gesneri Opera Botanica per duo saecula desiderata...cum figuris...ex bibliotheca D. Christophori Iacobi Trew...edidit D. C. C. Schmiedel. Norimbergiae impensis Io. Mich. Seligmanni. Typis Io. Iosephi Fleischmanni 1574. (*With separate title-page*) Valerii Cordi... stirpium...Liber Quintus...Editio nova...ex Gesneri Codice, 1751.

Schmiedel, C. C.
(1759–71)

Conradi Gesneri...Historiae Plantarum fasciculus quem ex bibliotheca Iacobi Trew...edidit et illustravit D. C. C. S. Norimbergiae, 1759.

Schreiber, W. L.
(1924)

Facsimileausgabe des Hortus Sanitatis, Deutsch, Peter Schoeffer, Mainz, 1485; Nachwort: Die Kräuterbücher des xv. und xvi. Jahrhunderts. München.

Schulz, A. (1919)

Euricius Cordus als botanischer Forscher und Lehrer. Abhandl. d. naturforschenden Gesellschaft zu Halle a.d.S. N.S. No. 7, 32 pp.

Appendix II

Schwertschlager, J. (1890)   Der botanische Garten der Fürstbischöfe von Eichstätt. Eichstadt.

Senn, G. (1923)   Das pharmazeutisch-botanische Buch in Theophrast's Pflanzenkunde. Verhandl. d. Schweizer. Naturforsch. Gesellsch., Zermatt, Teil II, pp. 201–2.

Senn, G. (1925)   Die Einführung des Art- und Gattungsbegriffs in die Biologie. Verhandl. d. Schweizer. Naturforsch. Gesellsch., Aarau, Teil II, pp. 183–4.

Seward, A. C. (1935)   The Foliage, Flowers and Fruit of Southwell Chapter House. Camb. Antiquarian Soc. Communications, vol. XXXV, pp. 1–32.

Sibthorp, J. (1806–40)   Flora Graeca. (Edited by Smith, Sir J. E., & Lindley, J.) London.

Sigerist, H. E.   See **Howald, E.**

Simler, J. (1566)   Vita...Conradi Gesneri...Tiguri excudebat Froschouerus.

Singer, C. (1927)   The Herbal in Antiquity and its Transmission to Later Ages. Journ. Hellenic Studies, vol. XLVII, pp. 1–52.

Sprague, T. A. (1928)   The Herbal of Otto Brunfels. Journ. Linn. Soc. Lond., Bot., vol. XLVIII, pp. 79–124.

Sprague, T. A. (1933¹)   Botanical Terms in Pliny's Natural History. Kew Bull., No. 1, pp. 30–41.

Sprague, T. A. (1933²)   Botanical Terms in Isidorus. Kew Bull., No. 8. pp. 401–7.

Sprague, T. A. (1933³)   Plant Morphology in Albertus Magnus. Kew Bull., No. 9, pp. 431–40.

Sprague, T. A. (1933⁴)   Botanical Terms in Albertus Magnus. Kew Bull., No. 9, pp. 440–59.

Sprague, T. A. (1936)   Technical Terms in Ruellius' Dioscorides. Kew Bull., No. 2, pp. 145–85.

Sprague, T. A. & Nelmes, E. (1931)   The Herbal of Leonhart Fuchs. Journ. Linn. Soc. Lond., Bot., vol. XLVIII, pp. 545–642.

Sprague, T. A. & Sprague, M. S. (forthcoming)   The Herbal of Valerius Cordus. Journ. Linn. Soc. Lond., Bot. [Abstr. Proc. Linn. Soc. Sess. 149, 1936–7, pp. 156–8].

Appendix II

Sprengel, K. P. J.
(1817, 18)

Geschichte der Botanik. Altenburg und Leipzig.

Sprengel, K. P. J.
(1822)

Theophrast's Naturgeschichte der Gewächse. Uebersetzt und erläutert von K. Sprengel. Altona.

Stadler, H. (1921)

Albertus Magnus de animalibus libri xxvi. Vol. II, books xiii–xxvi. (Beiträge z. Geschichte der Philosophie des Mittelalters, vol. xvi. Münster.)

Steinschneider, M.
(1891)

Die griechischen Aerzte in arabischen Uebersetzungen. § 30. Dioskorides. Virchow's Archiv für path. Anat., vol. cxxiv (ser. 12, vol. iv), pp. 480–3.

Stillman, J. M.
(1920)

Theophrastus Bombastus von Hohenheim called Paracelsus. Chicago.

Strömberg, R. (1937)

Theophrastea. Göteborgs k. Vet.- och Vit.-Samhälles Handl., F. v, ser. A, vol. 6, no. 4, 234 pp.

Strunz, F. (1903)

Theophrastus Paracelsus. Das Buch Paragranum. Herausgegeben und eingeleitet von Dr phil. F. S. Leipzig.

Strunz, F. (1926)

Albertus Magnus: Weisheit und Naturforschung im Mittelalter. Wien und Leipzig.

Stübler, E. (1928)

Leonhart Fuchs. München.

Thompson, D'Arcy W. (1913)

On Aristotle as a Biologist. Herbert Spencer Lecture. Oxford.

Toni, G. B. de
(1906)

Sull' origine degli erbarii. Atti d. Soc. d. nat. e mat. di Modena, ser. 4, vol. viii, anno xxxix, 1907 (for 1906), pp. 18–22.

Toni, G. B. de
(1907)

I Placiti di Luca Ghini...intorno a piante descritte nei Commentarii al Dioscoride di P. A. Mattioli. Venezia.

Treviranus, L. C.
(1855)

Die Anwendung des Holzschnittes zur bildlichen Darstellung von Pflanzen. Leipzig.

Trew, C. J. (1752)

Librorum botanicorum catalogi duo...Norimbergae Stanno Fleischmanniano.

Veendorp, H.
(1935)

Mededeelingen uit den Leidschen Hortus. II. De beteekenis van Charles de l'Escluse voor den Hortus te Leiden. (English summary.) Nederlandsch Kruidkundig Archief, pt. 45, pp. 25–96.

Appendix II

Ventura, A. F. G. (1933)   Clúsio. Portugal...nas suas obras. Coimbra.

Walton, I. (1670)   The Lives of Dr John Donne, Sir Henry Wotton, Mr Richard Hooker, Mr George Herbert. London.

Weber, F. P. (1893)   A Portrait Medal of Paracelsus on his death in 1541. The Numismatic Chronicle, ser. 3, vol. XIII, pp. 60–71.

Wegener, H. (1936)   Das grosse Bilderwerk des Carolus Clusius in der Preussischen Staatsbibliothek. Forschungen und Fortschritte, Berlin, Jahrg. 12, No. 29, pp. 374–6.

Wellmann, M. (1897)   Krateuas. Abhandl. d.k. Gesellsch. d. Wiss. zu Göttingen, Phil.-Hist. Klasse, N.S. vol. II, 1897–9, No. 1, 1897, pp. 1–32.

Wellmann, M. (1906–14)   Pedanii Dioscuridis Anazarbei de materia medica libri quinque. 3 vols. Berlin.

Wilms, H. (1933)   Albert the Great. (English version.) London.

Winckler, E. (1854)   Geschichte der Botanik. Frankfurt.

Winckler, L. (1934)   Das Dispensatorium des Valerius Cordus. Faksimile des im Jahre 1546...ersten Druckes durch Joh. Petreium in Nürnberg. Gesellsch. f. Geschichte der Pharmazie. Mittenwald (Bayern).

Wittrock, V. B. (1905)   Catalogus illustratus iconothecae botanicae Horti Bergiani Stockholmiensis. Pars II. Acti Horti Bergiani, vol. III.

Wolf, K. (1577)   Epistolarum medicinalium Conradi Gesneri... per C.W....in lucem data. Tiguri...Christoph. Frosch.

Wootton, A. C. (1910)   Chronicles of Pharmacy. London.

Zahn, G. (1901)   Das Herbar des Dr Caspar Ratzenberger (1598) in der Herzoglichen Bibliothek zu Gotha. Mitt. des Thüringischen Bot. Vereins, Weimar, N.S. Heft 16, pp. 50–121.

For a further discussion of some subjects treated in this book, see Arber, A. (1953), *From Medieval Herbalism to the Birth of Modern Botany*, in *Science Medicine and History*, Essays in honour of Charles Singer, ed. by Underwood, E. A., vol. I, pp. 317–36. Oxford.

APPENDIX III

SUBJECT INDEX TO APPENDIX II

[Under each heading, authors' names are given in alphabetical order. Dates are included only where an author is responsible for more than one memoir. It should be understood that the entries under individual names and subjects are merely supplementary to those under general headings, such as *History of botany, Incunabula*, etc. Some entries not in the general Index will be found here.]

Acosta: Olmedilla y Puig (1899); Ventura
Albertus Magnus: Fellner; Meyer & Jessen; Pouchet; Sprague (1933³), (1933⁴); Stadler; Strunz (1926); Wilms
Aldrovandi: Mattirolo (1897), (1898), (1899), (1907); Toni (1907²)
Amatus Lusitanus: Jorge
American plants: Emmart; Gabrieli; Killermann (1909)
Anguillara: Béguinot; Langkavel; Legré (1901)
Anne de Bretagne's Book of Hours: Camus (1894)
Apothecaries, Garden of Society of: Drewitt; Field
Apuleius Barbarus: see Apuleius Platonicus
Apuleius Platonicus: Cockayne; Gunther (1925); Howald & Sigerist; Hunger (1935²); see also under **Incunabula**
Aristotle: Hett; Lones; Strömberg; Thompson

Badianus herbal: Emmart; Gabrieli
Barnacle-goose: Heron-Allen; Stadler
Bartholomaeus Anglicus: see Incunabula
Bauhin, G.: Candolle; Christ (1913); Hess; Legré (1904); Savage (1935), (1937)
Bauhin, J.: Bauhin (1591); Legré (1904); Planchon
Belon: Legré (1901)
Besler: Schwertschlager
Bibliography, General: Copinger; Hain; Hunger (1917¹); Jackson (1881); Pritzel (1872, 7); Trew

Bock (Tragus): Adam; Irmisch (1859); Roth (1898)
Bohemia, History of botany in: Maiwald
Book of Hours of Anne de Bretagne: Camus (1894)
Book of Nature: see **Konrad von Megenberg**
Botanic gardens, History of, Hill (1915)
Bourdichon: Camus (1894); Savage (1923²)
British botany, General history of: Gunther (1922); Pulteney
British plants, First records of: Clarke. **History of names of:** Alcock; Britten (1881); Britten & Holland; Prior
Brosse: Arber (1913¹)
Brunfels: Arber (1921), (1936); Christ (1927); Church; Greene; Rytz (1933), (1936); Schmid; Sprague (1928); Wittrock
Burser: Savage (1935), (1937)
Busbecq: Forster; Kickx

Camerarius: Adam; Irmisch (1862)
Cherler: Legré (1904)
Cibo: Penzig
Circa instans: Camus (1886); Dorveaux; see also under **Manuscript herbals**
Clusius: see l'Écluse de
Colonna (Columna): Bellini; Faraglia
Constantin: Legré (1901)
Cordus, E.: Adam; Greene; Schulz
Cordus, V.: Adam; Greene; Irmisch (1862); Schmiedel (1751–4); Sprague, T. A. & M. S.; Winckler, L.
Cube: Pritzel (1846)

304

Appendix III

Cûnrat: see **Konrad von Megenberg**

Desmoulins: Planchon
Dioscorides: Berendes; Daubeny; Emmanuel; Gunther (1934); Hill (1937); Karabacek; Langkavel; Penzig; Sibthorp; Singer; Sprague (1936); Steinschneider; Wellmann (1906–14)
Dodoens: Adam; Courtois; Hunger (1917[1, 2]); Meerbeck; Morren, C. (1851); Morren & d'Avoine; Rooses (1882, 3)
Dorsten: Roth (1902)
Dourez: Legré (1904)
Dracaena: Schenck
Dürer: Killermann (1910), (1911)

Egenolph: Roth (1902); Schmid
Engraving, Copper: Kristeller
Engraving, Wood: Crane; Hatton; Kristeller; Muther; Treviranus

First records of British plants: Clarke
Fuchs: Adam; Arber (1928); Choate; Church; Greene; Hizlerus; Mohl; Roth (1899[1]); Sprague & Nelmes; Stübler
Fungi: Istvánffi

Gardens, History of Botanic: Hill (1915)
Gerard: Clarke; Jackson (1876)
Gesner: Adam; Bauhin (1591); Hanhart; Jardine; Milt; Planchon; Schmiedel (1751, 4), (1759, 71); Simler; Wolf
Ghini: Penzig; Toni (1907)
Goodyer: Gunther (1922); Kew & Powell
Goose tree: Heron-Allen; Stadler
Grand herbier: Camus (1886); see also under **Incunabula**
Grete herball, **Proof sheet of**: Gibson
Grew: Arber (1913[1])

Herbaria, History of: Camus (1895); Candolle; Druce; Kessler; Mattirolo (1899); Meusnier de Querlon (on Montaigne); Penzig (1905); Pitton de Tournefort (1694); Saint-Lager (1885[1]); Sarton (1931); Toni (1906)

Herbarium of **Apuleius Platonicus**: Hunger (1935[2]); see also under **Incunabula**
Herbarius, German and Latin: see under **Incunabula**
History of botany: Adam; Adanson; Alcock; Cuvier; Daubeny; Greene; Gunther (1922); Haller; Jessen; Maiwald; Meyer; Miall; Pitton de Tournefort; Pulteney; Sachs; Sarton; Sprengel; Winckler, E.
Hortus sanitatis: see under **Incunabula**
How: Gunther (1922)

Illustration, History of botanical: Choulant (1841, 1926), (1857, 1924); Crane; Haller; Hatton; Jackson (1906), (1924); Locy; Muther; Payne (1885), (1903), (1908), (1912); Savage (1923); Treviranus
Incunabula: Choulant (1841, 1926), (1857, 1924); Fischer; Hunger (1935[1]), (1935[2]); Klebs (1917, 8), (1925), (1926), (1932); Kommission f.d.G.d.W.; Payne (1885), (1903), (1908), (1912); Prideaux (1926); Saint-Lager (1885[2]); Schreiber
Isidorus: Sprague (1933[2])
Italian botany, General history of: Béguinot; Saccardo; Sprague & Nelmes

Johnson: Clarke; Gunther (1922); Kew & Powell; Ralph

Konrad von Megenberg: Locy; Pfeiffer; see also under **Incunabula**
Krateuas; Singer; Wellmann (1897)

L'Écluse, de: Boissard; Christ (1912); Cuvier; Hunger (1927); Irmisch (1862); Istvánffi; Legré (1901); Morren, C. (1853); Morren, É. (1875[1]); Planchon; Rooses (1882, 3); Roze; Veendorp; Ventura; Vorstius; Wegener
Le Moyne de Morgues: Savage (1922), (1923[2]), (1928, 9)
Linnean names of plants described in pre-Linnean herbals: Petermann, in Richter & Petermann

A H   *305*   20

Appendix III

L'Obel, de: Clarke; Gunther (1922); Legré (1899); Morren, É. (1875²); Planchon; Rooses (1882, 3)
Lonitzer: Adam; Roth (1902)
Lyte: Arber (1931); Downes; George; Maxwell-Lyte

Macer Floridus: Choulant (1832)
Manuscript herbals: Camus (1886), (1894); Cockayne; Choulant (1832); Fischer; Giacosa; Henslow; Hunger (1935²); Kaestner; Penzig; Saint-Lager (1885²) Sarton; Singer; see also under Apuleius, Dioscorides, Platearius
Mattioli: Fabiani; Savage (1921); Toni (1907); Walton (on Wotton)
Monardes: Gaselee; Olmedilla y Puig (1897)
Montpellier, School of botany at: Legré; Planchon
Musa: Howald & Sigerist

Names of British plants, History of: Alcock; Britten (1881); Britten & Holland; Prior
Nicolaus Damascenus: Hett (under Aristotle)

Orta: Markham; Olmedilla y Puig (? 1896); Ventura
Ortus sanitatis see under: Incunabula

Paracelsus: Adam; Hartmann; Stillman; Stoddart; Strunz (1903); Weber
Parkinson: Gunther (1922)
Pas or Passe, de: Savage (1923¹), (1928, 9)
Pena: Legré (1899); Planchon
Pharmacy, History of: Peters; Schelenz; Winckler, E. & L.; Wootton
Plantin: Degeorge: Rooses (1882, 3), (1909); Rooses & Denucé (1883–1918)
Plantin-Moretus Museum: Rooses (1909)
Platearius: Dorveaux; see also under Manuscript herbals
Platter: Adam; Arber (1936); Legré (1900¹); Meusnier de Querlon (Montaigne); Planchon; Rytz (1933), (1936)

Pliny: Brosig; Holland, P.; Sprague (1933¹)
Portuguese travel and botany: Jayne; Jorge; Ventura
Potato: Salaman
Proof sheet of Grete herball: Gibson
Provence, History of botany in: Legré (1899–1904)
Pseudo-Apuleius: see Apuleius Platonicus
Pseudo-Dioscorides: Kaestner

Rabel: Savage (1923¹, ²)
Ratzenberger: Kessler; Zahn
Rauwolff: Legré (1900²)
Raynaudet: Legré (1900²)
Records, First, of British plants: Clarke
Rhodion: see Rösslin
Rondelet: Planchon
Rösslin: Roth (1902)
Ruel: Sprague (1936)

Schongauer: Schenck
Southwell Chapter House carvings: Seward
Species and genus concepts, History of: Senn (1925)
Spenser and Lyte: Arber (1931)

Tabernaemontanus: Roth (1899²)
Terminology, History of: Choate; Sprague (1933¹), (1933³), (1936)
Thal: Irmisch (1862)
Theatrum Florae: Savage (1923¹)
Theodor: see Tabernaemontanus
Theophrastus: Greene; Hort; Senn (1923); Sprengel (1822); Strömberg
Thurneisser: Moehsen
Tragus: see Bock
Turner: Britten; Clarke; Jackson (1877)

Vérard: Macfarlane
Vettonica: Howald & Sigerist

Weiditz: Arber (1936); Rytz (1933), (1936)
Wood-engraving: see Engraving, Wood
Wormwood, Modern use of: Broadwood

Zaluziansky: Maiwald

306

INDEX

Index

άνθος, 155
"Antirrhinon", see Snapdragon
Antwerp, 79, 80, 82, 124, 215
"Apios", 213 (fig. 102)
"Apium ranarum", 149
"Apocynum", 99 (fig. 48)
Aprill eclogue (Spenser), 126, 127
Apuleius (Barbarus or Platonicus), see *Herbarium* (Apuleius Platonicus)
Aquilegia vulgaris L. (Columbine), 127, 147, 203, 204 (pl. xxi)
Aquinas, St Thomas, 4
Arabia, 4; A. minor, 25
Arabic commentators and translators of the Classics, 4, 12, 20, 104, 270
"Arabist", 104
Arbolayre, 26
"Arbor...scientie" (Tree-of-knowledge), 32, 33 (fig. 14), 200
"Arbor tristis", 102, 103 (fig. 50)
Arbutus Unedo L., 137
Arctium Lappa L., 208 (fig. 99)
Argemone, 179
Aristolochia ("Aristologia"), 10, 19 (fig. 4), 47
Aristotelian botany, 2–6, 108, **143**–**145**–147, 163, 164; see also Theophrastus
Aristotle, 2, 3, 4, 23, 143, 272
"Armes parlantes" of M. de l'Obel, 89
Arnaldus de Villa Nova, 22, 273; see also *Tractatus de virtutibus herbarum*
"Aromata", 179
Arras, 85
Arrowhead, 90
Art of Simpling, The (Cole), 254, 285
Artas, 40
Artemisia (Mugwort, Wormwood), 17 (fig. 2), 39, 47, 49, 59, 262, 263
Arum maculatum L. (Wild-arum), 50, 177, 216 (fig. 104), 217
Arzt in seinem Studierzimmer, Der (Ostade), Frontispiece, 91
As You Like It, 39
Asarum europaeum L. (Asarabacca), 205 (fig. 97)

Asia, 55; A. minor, 8
Askham's *Herball*, 43, 274
Asparagus officinalis L., 166 (fig. 71), 175 (fig. 78), 210, 217
Assafoetida, 105
Astragalus Tragacantha L., 173
Astrology, Botanical, 247, 252, **256**–**263**, 266
Athos Peninsula, 11
Atropa Belladonna L., 90
Augsburg, 14, 19, 194
Austria, 87, 89
"Avena", 95 (fig 45)
Avicenna, 22, 25, 248, 249
Avium praecipuarum (W. Turner), 133 (footnote)
Aztec medicine, 109, 110

Babylonia, 25
Badianus, J., 110
Baldo, Viaggio di Monte, see *Viaggio di Monte Baldo*
Bamboo, 217
Bämler, H., 14
Banckes' *Herball*, 41–44, 149, 274
"Barbary bush", 137 (fig. 61)
Barnacle goose, see "Barnakle tree"
"Barnakle tree", 130, 131 (fig. 60)–134
"Barnakles, Breede of", see "Barnakle tree"
Bartholomaeus Anglicus, 2, **13, 14,** 41, 42 (fig. 19), 190, 271
Bartholomew de Glanville, 13, 271
Basilisk, 34
Basing House, Siege of, 135
Basle, 64, 113, 114, 248
Bassaeus, N., 76
Bassano, L., 102 (pl. xi)
Bath and Wells, Bishop of, 120
Batthyány, Baron B. de, 88, 266
Bauhin, G. (C.), 93, 113, **114** (portrait, pl. xiii), 115 (fig. 55), **116,** 118, 134, 137, **159** (fig. 68), **160,** **168, 179, 181,** 237, 266, 268, 281–283, 285
Bauhin, J., the Elder, 113; the Younger, 70, 85, 93, **113, 114,** 115, 118, 119, 283
"Bauser vel Bausor", 31 (fig. 12), 32

308

Index

Bavaria, 20, 64
"Baye tree", 257
Beatrice, 258
Bedfordshire, 270
Belch, Sir Toby, 23
"Bell-form" of flower (Albertus), 147
Bellenden, J., 133 (footnote)
Belon, P., 234, 235 (fig. 120), 276; see also *Les Observations*
"Benedictenwurzel", 57 (fig. 25)
Berberis ("Barbary bush"), 137 (fig. 61), 237
Bergian Library, 52 (pl. vi)
Bergzabern, Dietrich of, see Tabernaemontanus
Bern, 52, 141, 206
Bernicles, see "Barnakle tree"
Besler, B., 244, 282
"Beta Cretica semine aculeato", 159 (fig. 68), 160; "B. nigra", 160
Betel-nut, 105
Betony, 188
Binary system of nomenclature, 116, **168**
Biophytum sensitivum Dec., 105
"Bird-form" of flower (Albertus), 147
Bird-lime, 152
Birds, Turner on, 124, 132, 133
Bird's-nest-orchid, 177
Bittersweet, 148 (fig. 62)
Black background in woodcuts, 189, 190, 200, 209
Black-bryony, 177
Blackfriars, 236
Black-hellebore, 7, 8
Bladder-nut, 62 (fig. 28)
Bock, J. (Tragus, H.), 38, 52, **55**, **58** (portrait, fig. 26), **59**, 60 (fig. 27), **61**, 62 (fig. 28), 63 (fig. 29), 64, 67, 70, 76, 122, **151–153**, 160, **166**, 219, 220 (fig. 106), 221 (fig. 107), 222 (fig. 108), 266, 268, 275
Boethius (Boece), H., 133
Bohemia, 144, 181
Boke of Distyllacyon, see *Distyllacyon, Boke of*
Boke of secretes of Albartus Magnus, 256, 257, 272

Bologna, 10, 75, 77, 114, 120
Bombast von Hohenheim, T., see Paracelsus
Bonhome, M., 203 (fig. 95)
Bonony, see Bologna
Book of Nature (Konrad), 14 (pl. iii), 189, 190, 267, 271
Borage, 147
Borax, 49
Borcht, P. van der, 229, 230
Botanic garden, Earliest, 100
Botanicon (Dorsten), 72, 73 (title-page, fig. 36)
Botanologia (R. Turner), 254, 255, 285
Botany School, Cambridge, 110 (pl. xii), 144 (pl. xvii)
"Botris", 199 (fig. 93)
Botrychium, 221 (fig. 107)
Boussuet, F., 203 (fig. 95), 204
Box, 152
"Brake", 61
Bramble, 41
Brasavola, A., 11, 275
"*Brassicae quartum genus*", 66 (fig. 31)
Braunfels, 52
Braunschweig, see Jerome of Brunswick
Bredwell, S., 129
"Breede of Barnacles", see "Barnakle tree"
"Breyt wegrich", 180 (fig. 80)
Brimen, Marie de, 89
"Brionia" (Bryony), 21 (fig. 6), 190 (fig. 85)–192
"British Botany, Father of", 119
British Museum, 35, 38, 74, 76 (pl. vii), 90 (pl. ix), 125, 188, 189, 236
British plants, First records, and study of localities, 90, 124, 134, 137
Broomrape (*Orobanche*), 124, 177
Brosse, G. de la, 144, 145, 250, 255, 284
Browne, Sir T., 244
Brunfels (Brunfelsius), O., 26, **52** (portrait, pl. vi), 53 (fig. 22), 54 (fig. 23), **55**, 56 (fig. 24), 57 (fig. 25), 58, **61**, 67, 70, 76, 149, 150, 151, 152, 170 (fig. 75), **202**, 203,

309

Index

Index

Index

312

Index

Index

314

Index

Index

Index

317

Index

Index

319

Index

Montpellier, 85, 90, 91, 113, 114, 118, 119
Moonwort, 221 (fig. 107)
More, H., 254
Moretus, E., 80; J., 80
Morison, R., 116, 138
Morpeth, 119
Morus nigra L. (Mulberry), 224, 226 (fig. 112)
Mosses, 181
Mount Sinai, 25
Moyne, J. Le, see Le Moyne de Morgues, J.
Mugwort, see Artemisia
Mulberry, see Morus nigra L.
Munich, 14
"Musci", 182
Musée Plantin-Moretus, 80, 81
Museum of natural history, First, 102
Mushrooms ("Mussherons"), 153, 162, 171, 263
Musk, 48
Mutability of species, 5
My mynde to me a kyngdome is, 108
Mycology, Founder of, 88
Myrtle, 166

Names of herbes (W. Turner), 121, 276
Naples, 44, 97, 243, 251
Narcissus, 32 (fig. 13), 89
Nassau-Saarbrücken, Philip, Count of, 59
Native herbs, Advantages of using, 255, 256, 261
Natura stirpium, De, see De Natura stirpium (Ruellius)
Natural History (Pliny), 8, 12, 14, 165, 166, 188, 234
Nature... des plantes, De la (Brosse), 144, 145, 250, 255
Natures and Elements, Four, 23, 24, 45, 46
Nederlandste Herbarius (Nylandt), 91, 92, 286
"Nenufar", 49 (fig. 21), 149, 169 (fig. 74)
"Nenuphar", 170 (fig. 75)
Neottia, 177
Nero, 8

Netherlands, see Low Countries
"Nettel", see Stinging-nettle
Neuw Kreuterbuch (Tabernaemontanus), 76, 115, 281
Neuw vollkommentlich Kreuterbuch (Tabernaemontanus), 281
New College, Oxford, 252
New Herball, A (W. Turner), 70, 121–124, 276
New Herball of Macer, A, 44, 274
New Kreüterbüch (Fuchs), 67, 153, 162, 212, 275
New Kreütter Buch, see Kreütter Buch, New (Bock)
Nicolai, A., 229
Nicolaus Damascenus, 4
Nicotiana, 179; N. Tabacum L., 109 (fig. 53), 229, 230 (fig. 116)
Nievve Herball, A (Dodoens and Lyte), 70, 125, 126 (fig. 58), 127, 128 (fig. 59), 155, 277
Nigella, 176
"Ninfea", 174 (fig. 77), 241
Noah, 136
Noir, P. le, 36
Nomenclature, Binary system of, 116, 168
Norman Conquest, 38, 41
Northumberland, 119
Norton, G., 76, 129
Nova stirpium adversaria (Pena and de l'Obel), 278
Nuphar, 177; N. luteum Sm., 172 (fig. 76), 241; see also Water-lily
Nuremberg, 75, 77; N. Dispensatorium, 75
Nutmeg, 105
Nylandt, P., 70, 91, 92, 285
"Nymfea", 167 (fig. 72), 239
Nymphaea, 177, 241; N. alba L., 170 (fig. 75); see also Water-lily
"Nymphea", 167 (fig. 73)

Oak, 151 (fig. 64), 166, 225 (fig. 111)
"Ocimoides fruticosum", 77 (fig. 37)
Octavian, the Irishman, 133
Odo, 44, 272
Officina Plantiniana, 229

320

Index

Index

Phytognomonica (Porta), 204 (fig. 96), 236, 237, 251, 252, 253 (fig. 127), 256, 257 (fig. 128), 281
Phytography, see Description
Phytologia Britannica (How), 285
Phytopinax (G. Bauhin), 116, 168, 281
"Pile of Foulders", 132
"Pimienta negra", 106 (fig. 51)
Pimpernel (Anagallis), 10, 155
"Pimpernuss", 62 (fig. 28)
Pinax (G. Bauhin), 116, 137, 179, 268, 283
Pinet, A. du, 119
"Pinkes", 128 (fig. 59), 135
"Pionia", 150 (fig. 63)
"Piper Nigrum", 106 (fig. 51)
Pisa, 64, 77, 144 (pl. xvii)
Pistil (Pistillum), 152
Pitton de Tournefort, J., 116, 142
Plague, 64, 92, 93, 111
Plantaginaceae, 176
Plantago (Plantain), 15 (fig. 1), 177, 188, 257; "P. major", 180 (fig. 80); "P. quinquenervia rosea", 161 (fig. 69)
Plantarum...icones (de l'Obel), 91, 176, 280
Plantarum seu stirpium historia (de l'Obel), 91, 278
Plantin, C., 79–82, 85, 86, 91, 118, 134, 176, 215, 228, 229, 232, 233, 239, 241, 266
Plantin, Maison, see Musée Plantin-Moretus
Plantis Aegypti, De, see De plantis Aegypti (Alpino)
Plantis, De (Nicolaus Damascenus), 4
Platearius, 25, 26
Plato, 2
Platter, F., 85, 114, 141, 206
Pliny, The Elder, 8, 12, 14, 165, 166, 188, 234
"Polygonum latifolium" (P. multiflorum All.), 68 (fig. 32)
Pomegranate, 252
Poplar, White, 89
Popp, J., 252, 284
Poppy, 61, 147; Welsh, 137

Porta, G. B., 204 (fig. 96), 236, 237, 251, 252, 253 (fig. 127), 255–257 (fig. 128), 281
Portraiture by herbal artists, 217, 219, 243
Portugal, 86, 104, 105; see also Lisbon
Portuguese India, 104, 105
Potato, 89, 115 (fig. 55), 116, 130 (pl. xiv), 157, 158 (fig. 67)
Prague, 95, 144, 223, 224
"Prestes hode", 50
Preussische Staatsbibliothek, Berlin, 229, 230 (pl. xxii)
Priest, Dr, 129
Primrose, 49
Principles, Four, 23, 24, 45, 46
Printing, Invention of, 13, 20
Prodromos theatri botanici (G. Bauhin), 115 (fig. 55), 116, 159 (fig. 68), 160, 237, 268, 283
Professorship of Botany, Earliest, 100
Proof sheet of Grete herball, 44, 45
Proportion, disregard of, in early illustrations, 193
Prunus persica Stokes, see Peach-tree
Pryme, A. de la, 138
"Prymerolles", 49
Pseudo-Apuleius, see Herbarium (Apuleius Platonicus)
"Psyche" in plants, 2, 5
"Ptysan" of barley water, 47
Pûch der natur, Das (Konrad), 14 (pl. iii), 189, 190, 271
"Pulsatilla", 183 (fig. 82)
Pumpkin-gourd, 69 (fig. 33), 70
Purgantium...historiae (Dodoens), 279
"Pynkes", see "Pinkes"
"Pyra", 94 (fig. 44)
"Pyramid-form" of flower (Albertus), 147
Pyrola, 220 (fig. 106)
Pyrus communis L., 94 (fig. 44); P. Malus L., 98 (fig. 47)

Quakelbeen, W., 97
Quatro libros...Nueva España (Hernandez), 109, 283

322

Index

Index

Index

Index

Van der Loe (Vanderloe), J., 82, 227, 229
Vatican Library, 110
Venice, Venetian republic, 92, 95, 100, 223
Vérard, A., 35, 74
Vervain, 47
Vespasian, 8
"Vetonica altilis" (Pinks), 128 (fig. 59)
"Vettonica" (Betony), 188
Viaggio di Monte Baldo (Calzolari), 100, 279, 280
Victoria and Albert Museum, 236
Vienna, 9, 14, 82, 86–88, 204 (pl. xxi)
Vienna Codex (Dioscorides), see *Codex Aniciae Julianae*
Vigna sinensis Endl. or *V. unguiculata* (L.) Walp., 186 (pl. xviii)
Vinci, L. da, 202 (pl. xx), 203, 217
Vine, 5, 153, 162, 192
Violet, 147, 189
Virginia, 130
Viribus herbarum, De (Macer), 44, 272
Virorum doctorum effigies, 93 (fig. 43)
Virtutibus herbarum, De, 6, 256, 272
Viscum album L. (Mistletoe), 51, 151 (fig. 64), 152, 229
Vita Conradi Gesneri (Simler), 111, 112 (fig. 54)

Walnut, 252, 253
"Walwurtz männlin", 53 (fig. 22)
War, Civil, The, 135
Watènes, Seigneur de, 86
"Water crowfote", 149
Water-lily, 16, 49 (fig. 21), 149, 167 (figs. 72, 73), 169 (fig. 74), 170 (fig. 75), 172 (fig. 76), 174 (fig. 77), 177, 188, 193, 217, 229, **239, 241**
Water plants, "Swimming", 177
"Waterworte", 41
"Wegrich, Breyt", 180 (fig. 80)
Weiditz, H., 55, 206, 207, 209, 268; see also Brunfels, O.
Wells, Deanery of, 120
Welsh-poppy, 137
Wemding, 64
Westminster, Garden in, 135
White-bryony, 40

White-poplar, 89
White-water-lily, see *Nymphaea alba* L.
Whorls, Alternation of, in the flower, 147
"Whyte elebore", 48
Wild-arum, see *Arum maculatum* L.
William the Silent, 90
Willows, Theophrastus on, 166; "wool" of, 61
Winckler, N., 258, 279
Windisch land, 25
Windsor, Royal Library, 202 (pl. xx)
Winter-aconite, see *Eranthis hiemalis* L.
Winter-cherry, 194, 196 (fig. 90), 197 (fig. 91), 198, 199
"Winter gardens", see Herbaria
Winter-green ("Winter grün"), 220 (fig. 106)
Wittenberg, 74, 77, 140
Wolf, 257
Wolf, K., 111
Wonder Trees (Centuria Arborum Mirabilium), Olorinus, 18, 283
Woodcuts (wood-engravings), botanical, 186–241, *et passim*; graver's burin, 225, 226 (figs. 112, 113)
Woodpecker, 7
Wood-sorrel, 50, 148, 149, 177
Wormwood, see *Artemisia*
Worthy practise, A (Fuchs), 64
Wotton, Sir H., 95–97, 135
Wyer, R., 44
Wynkyn de Worde, 41, 190

Ximenez, F., 283

Yellow-flag, see *Iris Pseudacorus* L.
"Yvery" (Ivory), 46 (fig. 20)

Zaluziansky, A., 93, **144, 181, 182,** 183, 281
Zea Mays L. (Indian-corn), 70
Zocchi, G., 144 (pl. xvii)
Zodiac, Signs of, 257, 260, (fig. 129, pl. xxv), 261–263
Zouche, Lord, 90
"Zparagus", 166, (fig. 71)
Zurich, 110, 111, 219, 265
Zweibrücken, 59